Conquer Math.
Conquer Fear.

Conquer Math. Conquer Fear.

The Focus-Driven Playbook for Turning Anxiety Into Math Achievement

by

Dr. Joe Evans

ISBN (paperback): 979-8-9951630-1-5

First edition: 2026
Printed in the United States of America
Cover design: Bria Noel Evans
Interior design: Mitch Stimers, Compita Consulting
Editors: Mitch Stimers, Compita Consulting, Kelly Evans
and Dr. Joe Evans, Evans Entertainment Group

Acknowledgements

First and foremost, I give all honor, glory, and praise to God. Thank You for blessing me with the wisdom, strength, perseverance, and purpose to complete this work. Your grace has carried me through every challenge, and without You, this book would not have been possible. I pray that every reader is encouraged to approach mathematics with confidence, courage, and a renewed belief in their ability to succeed.

To my incredible wife, Kelly Evans, thank you for being my greatest supporter and trusted editor. Your careful attention to detail, thoughtful feedback, and unwavering encouragement strengthened every chapter of this book. More importantly, your love, patience, and belief in me have been a constant source of inspiration throughout this journey. I could not have done this without you.

To my talented daughter, Bria Noel Evans, thank you for designing the book **cover**. Your creativity transformed the vision of *Conquer Math, Conquer Fear* into something that captures its heart and purpose. I am incredibly proud of your artistic gifts and grateful for your contribution to this project.

I also extend my deepest appreciation to the countless teachers, students, families, mentors, and colleagues who have shaped my understanding of mathematics and education over the years. You have taught me to appreciate not only the beauty of mathematics but also its complexity, logic, creativity, and endless possibilities. Every classroom conversation, every challenge, every success, and every opportunity to learn alongside you has influenced the ideas shared in these pages.

Finally, to every student who has ever questioned whether they were "good at math," this book is for you. Always remember that confidence can be taught, fear can be overcome, and with the right mindset, strategies, and perseverance, every learner can achieve more than they ever imagined.

About the Book

Conquer Math, Conquer Fear: The Focus-Driven Playbook for Turning Math Anxiety into Achievement empowers students to overcome the fear that often stands between them and mathematical success. Drawing from decades of experience as a mathematics educator, instructional leader, principal, mentor, and tutoring provider, Dr. Joe Evans shares practical, research-informed strategies that build confidence, strengthen problem-solving skills, and foster a growth mindset. Through real-life math examples, proven techniques, and a focus-driven approach, this book demonstrates that every student can transform math anxiety into resilience, confidence, and lasting achievement.

Preface

As an adolescent, especially in middle school, math felt like a foreign language written in symbols I had never been taught to speak. I would sit at my desk staring at what looked like a chaotic swarm of numbers scattered across the page. I knew they were supposed to follow a pattern. I knew there was an order, a logic, a hidden rhythm to it all. But to me, it felt unorganized, overwhelming, and impossible to decode.

The "order of operations" might as well have been an encrypted formula. Multiply, divide, add, subtract; everyone else seemed to move through it with quiet confidence while I struggled to make sense of where to even begin. I couldn't connect the dots. I couldn't process the steps. I couldn't see how one line led to the next. And because I didn't understand it, I feared it. That fear became louder than the numbers themselves.

I began to believe that maybe math just wasn't for me. Maybe my mind wasn't wired that way. Each unsolved problem felt less like a puzzle and more like proof of limitation. But what I didn't yet realize was this: the problem wasn't my ability; it was my belief. The numbers weren't unorganized. They were waiting for me to learn their language. And once I did, everything would begin to change.

This book is not just about math.

It is about fear.

It is about focus.

It is about the quiet lies we tell ourselves when we encounter something difficult, and the powerful truths we discover when we decide to confront it.

For years, I thought math anxiety was a personal weakness. I later learned it is one of the most common academic fears in the world. Brilliant students freeze at the sight of equations. Capable adults still avoid numbers whenever possible. Fear distorts clarity. Anxiety disrupts logic. Doubt blocks execution. The issue is rarely intelligence. It is mindset, strategy, and focus.

And that is exactly what this playbook is designed to rebuild.

Conquer Math Conquer Fear is a focus-driven blueprint for transforming anxiety into achievement. It is practical. It is psychological. It is strategic. And most importantly, it is personal. I have lived the frustration. I have felt the embarrassment. I have wrestled with the self-doubt. But I have also discovered the system that changes everything.

Here is what I learned:

• Fear thrives in confusion.

• Confusion shrinks with clarity.

• Clarity grows with focus.

• Focus produces confidence.

Confidence is not magic. It is built.

Inside these pages, you will learn how to:

• Rewire your mindset about math and intelligence.

• Break complex problems into manageable steps.

• Master focus techniques that calm anxiety instantly.

• Build mathematical confidence through structured repetition.

• Replace avoidance with action.

• Turn mistakes into momentum.

This is not about becoming a mathematician overnight. It is about becoming disciplined in your thinking. It is about learning how to stay present when your mind wants to panic. It is about developing a method you can apply not only to math, but to life.

Because the truth is this: math is not just numbers. It is structure. It is order. It is problem-solving. It is resilience in numeric form. When you learn to solve for *x*, you also learn how to solve for obstacles, setbacks, and uncertainty.

The principles in this book extend far beyond the classroom. Focus is transferable. Discipline is transferable. Confidence is transferable.

When you conquer math, you are not just passing a test.

You are proving to yourself that fear does not control you.

And once you learn that lesson, there is no equation too difficult, no challenge too intimidating, and no goal beyond your reach.

If you have ever stared at a problem and felt overwhelmed, this book is for you.
If you have ever doubted your ability to understand numbers, this book is for you.

If you are ready to replace anxiety with achievement, this book is for you.

You are not bad at math.
You were never incapable.
You simply needed a strategy.

Let's begin.

Quick-Reference Guide: One Strategy Per Chapter

#	Chapter	Core Strategy
1	Math Is Not the Enemy	Struggle is data, not a verdict on ability; name the fear cycle to start breaking it.
2	Rewriting the Narrative	Swap fixed-mindset language for growth-mindset language; practice reshapes the brain at any age.
3	Math Has a Language	Treat math vocabulary as a second language, using tools like the Frayer Model to master new terms.
4	Organization Is a Math Superpower	Build the systems, binders, note routines, trackers, that free the brain to focus on solving problems.
5	The Power of Note Taking	Turn note taking into an active thinking tool instead of a copying task.
6	Learning the Process, Not Just the Answer	Explain reasoning out loud; understanding beats memorizing an answer.
7	Focus Over Frustration	Train attention like a skill, with short routines that reset focus before frustration takes over.

8	Mistakes Are Part of Mastery	Treat every mistake as information to analyze, never as a verdict on ability.
9	Peer Pressure and Performance	Build belonging and representation so visible effort stops carrying a social cost.
10	Teachers Who Transform Math Culture	Pair high expectations with real support, trust, and culturally responsive teaching.
11	Families, Mentors, and Math Success	Involve family and mentors in ways that build confidence without requiring math expertise.
12	The Mathematics of Discipline	Turn consistency into a system: spaced practice, written goals, a schedule, and an accountability partner.
13	Real-Life Math and Future Opportunities	Connect math to real careers, money, and goals to build lasting motivation.
14	Building the Confident Math Student	Rehearse specific confidence practices: affirmation, expressive writing, and a fearless-response routine.
15	Conquer Math. Conquer Fear.	Draw every strategy into one personal blueprint to carry forward.

Full chapter detail, research citations, and reflection questions appear in the complete book. This page is meant to be posted, copied, and kept close at hand.

TABLE OF CONTENTS

Introduction

To the one holding this book: you did not open these pages because you are bad at math. You opened them because, somewhere along the way, math started to feel like an enemy rather than a subject, and you are ready for that to change.

I have spent years watching capable students, students with sharp minds and real potential, talk themselves out of a subject before they ever gave it a fair chance. Not because the math itself was too hard, but because fear got there first: fear of a wrong answer spoken out loud, fear of a stereotype they never asked to carry, fear that struggle meant something was wrong with them. None of that fear was ever a measure of ability. Putting that fear into words gives a student something to return to on the hardest days, and that is the reason this book exists.

Every strategy in the following 15 chapters comes from real research, named and cited, not from guesswork or good intentions alone. Confidence is not born ready-made; it gets one deliberate practice at a time. Teachers, families, and mentors carry real weight in that process, which is why this book speaks to each of them as well as to students.

Math will not become easy just because you own this book. Nothing worth building ever is. What this book offers instead is a plan, grounded in evidence, for facing a hard subject without fear standing in the way. Read it one chapter at a time. Practice what it teaches. Return to it when the fear tries to come back, because it will try.

Conquer math. Conquer fear. That work starts on the next page.

Math anxiety can produce measurable, documented interference with working memory and performance (Ashcraft, 2002; Steele & Aronson, 1995). None of the 15 chapters that follow treats that student as broken or behind; each chapter treats that student as capable, and each chapter names a specific, research-grounded strategy for closing the gap between where that student stands today and where their actual ability could take them. The belief that ability grows through effort and strategy, rather than being fixed at birth, changed how students responded to setbacks throughout their school careers, and that belief underlies every chapter in this book (Dweck, 2006). The 15 chapters move in a deliberate order: mindset and identity first, then the practical systems of vocabulary, organization, note taking, and reasoning, then focus, error analysis, and the social pressures of peer and classroom culture, then the adults surrounding a student at school and at home, then discipline, purpose, confidence, and a closing blueprint for carrying every earlier chapter forward. Family involvement, culturally responsive teaching, and community mentoring each appear as separate chapters because none of this work happens in a single student's head (Gay, 2000; Henderson & Mapp, 2002). The References and Additional Resources sections at the end of this book, together with this introduction, frame the 15 chapters as a single connected argument rather than as 15 separate essays.

How to Use This Book

Read the 15 chapters in order, since later chapters build on strategies and language introduced earlier, and the book's own transition paragraphs are written to maintain that continuity. Each chapter follows the same structure: a Chapter Purpose and Guiding Question, Learning Objectives, a 6-part Section Outline, a list of Research Foundations, Supplemental Features including a Math Mindset Moment and a Student Success Story, and a Chapter Takeaway that transitions into the next chapter. Three companion documents extend this book beyond its 15 chapters: a Teacher's Guide that translates each chapter into specific classroom practice, a Family and Mentor Guide that translates each chapter into specific home support, and a Glossary that collects all recurring terms in a single alphabetized reference.

(a) Read chapters in sequence, since each chapter's vocabulary and strategies build on chapters that came before it;

(b) use the companion Glossary anytime a familiar term reappears without its full original explanation nearby;

(c) pair a student's reading of each chapter with the matching entry in the Teacher's Guide or the Family and Mentor Guide, depending on who is supporting that student.

Chapter 1: Math Is Not the Enemy

Breaking the Fear Cycle in Mathematics

Getting Started

Many students grow up believing that mathematics is their greatest academic obstacle, and they come to see math as something to fear rather than understand. A poor test grade, a difficult homework assignment, or even a careless comment like, “I’ve never been good at math,” can shape a student’s confidence for years. The truth is that mathematics is not the enemy. Fear, self-doubt, and negative thinking are the real barriers to success. Just as an athlete cannot win a game while already believing they will lose before stepping onto the field, students cannot expect to succeed in mathematics if they have already convinced themselves that failure is inevitable. The first step toward conquering math is changing the way we think about it.

The reality is that mathematics is part of our everyday lives. Every time you purchase a new pair of shoes during a 30% off sale, calculate the amount of money needed for a family vacation, compare cell phone plans, or determine how long it will take to drive to a destination, you are using mathematical thinking. Imagine buying a video game that costs $69.99 with a 15% discount. Understanding percentages allows you to quickly calculate the savings before reaching the checkout. Professional athletes also depend on mathematics. A basketball player studies shooting percentages to improve performance, while a baseball manager analyzes batting averages and earned run averages to make strategic decisions. Math is not confined to a classroom: it is a life skill that influences nearly every decision we make.

Throughout this book, you will discover that mathematics is less about memorizing formulas and more about developing confidence, perseverance, and problem-solving skills. Every successful engineer, nurse, architect, entrepreneur, teacher, and scientist once struggled with difficult problems, succeeding because they learned to embrace challenges rather than avoid them. Near the end of this journey, you will begin to see mathematics not as an enemy standing in your way, but as a trusted companion helping you build the future you desire.

Real-Life Math Connection

Imagine your favorite pair of sneakers costs $120, but the store is offering 25% off. How much will you save?

Solution

Discount = 25% × $120

= 0.25 × 120

= $30

Final Price = $120 − $30 = $90

Understanding percentages allows you to know whether you're getting a good deal before reaching the register. Every smart shopper uses math.

Chapter Purpose and Guiding Question

How does fear, rather than lack of ability, shape a student's relationship with mathematics?

Chapter 1 establishes the emotional and psychological foundation for the entire book: fear, not aptitude, drives the widest gap between students and mathematical success. Classrooms across the country host students who mask unresolved confusion behind practiced indifference, choosing silence over the risk of a wrong answer spoken aloud. Embarrassment operates as a gatekeeper in these rooms, closing off questions before a hand ever rises. Peer pressure compounds the problem when academic effort reads as a betrayal of social identity rather than a personal strength. Negative self-beliefs, once formed, harden into a script that students rehearse long after the original wound has faded. Stereotype threat research demonstrated that students underperformed on tests framed as measures of intellectual ability when they were aware of a negative stereotype about their group, even when their actual skill matched that of peers outside the stereotyped group (Steele & Aronson, 1995). Chapter 1 opens with this evidence to name a cycle many students have lived without ever seeing it labeled. Naming the cycle gives students and teachers a shared vocabulary for the work ahead. The remaining sections trace that cycle from its emotional roots to the first steps toward a growth-oriented identity, preparing readers for the mindset work of Chapter 2.

Learning Objectives

Chapter 1 sets five learning outcomes for students to reach before its final page:

(a) name the fear cycle and identify at least one personal experience that fits its pattern;

(b) distinguish between disliking math and fearing judgment for struggling with math;

(c) define math anxiety and describe two of its documented effects on working memory and test performance (Ashcraft, 2002);

(d) explain, in their own words, why struggling with a math problem does not indicate low intelligence;

(e) identify one growth mindset strategy they can practice during the next math lesson (Dweck, 2006).

Section Outline

I. The Fear Cycle: Definition and Origins

The fear cycle begins long before a student opens a math textbook, and it seldom announces itself with a label attached. Early experiences of public correction or confusion attach a lasting emotional charge to the subject, one that resurfaces whenever new material appears. Repeated exposure to that charge trains a student to expect failure in advance, a pattern tied to reduced persistence on unfamiliar problems (Ashcraft, 2002). Fear, once established, narrows a student's willingness to take the high risks that mathematical learning requires. A guided walk through the anatomy of the cycle closes this section, giving students language to catch the pattern in real time.

(a) The trigger event: a public mistake, a timed test, or a moment of comparison with a peer;

(b) the emotional response: shame, embarrassment, or a spike in anxiety that narrows working memory;

(c) the avoidance behavior: disengagement, guessing, or refusal to attempt new problems;

(d) the reinforcing belief: "I am not a math person," a conclusion that closes the cycle and sets up its repetition.

II. The Psychological and Cognitive Toll of Math Anxiety

Math anxiety operates as more than discomfort; it consumes cognitive resources a student needs to solve the problem in front of them. Working memory carries the load of holding numbers, steps, and strategies in mind at once, and anxiety competes for that same capacity. Students with pronounced math anxiety showed measurable declines in accuracy and speed on tasks that required heavy working memory use, even when their math knowledge matched that of their less anxious peers (Ashcraft, 2002). The cost extends past the classroom: avoidance of math courses, majors, and careers followed students who never resolved their early anxiety.

(a) Reduced working-memory capacity during timed or high-pressure tasks;

(b) increased avoidance of math courses, electives, and math-heavy career paths;

(c) physical symptoms such as a racing heart, tense shoulders, or a blank mind during tests (Ashcraft, 2002).

III. The Mask of Indifference: Why Students Perform Dislike

Many students announce that they hate math long before they explain why, and the declaration often protects them from a harder admission: they are afraid. A performed dislike costs less than a visible attempt that ends in a wrong answer in front of classmates. Indifference reads as a choice, while a wrong answer reads as a limit, and most students protect their sense of self through the version they control. Teachers who mistake the mask for the truth respond with lower expectations, which then confirms the very belief the mask was built to hide.

(a) Verbal dismissal ("math is boring," "I don't need this") used as a shield before an answer is even attempted;

(b) social costs of visible effort, including teasing or accusations of "acting smart";

(c) The gap between private capability and public performance, a gap that widens each time the mask goes unchallenged.

IV. Peer Pressure and the Politics of Visible Intelligence

Classrooms carry social rules that operate alongside academic ones, and those rules can punish visible intelligence among students who fear standing apart from a peer group. Academic effort is read as a betrayal of group identity for some students, a social cost that shapes real classroom behavior even when no adult in the room ever names it. School influence, community forces, and individual agency combine to shape a student's math identity and outcomes across a wide range of settings. A student weighing that social cost against an academic reward will often choose the peer group, not because the reward lacks value but because the cost feels immediate and personal. Chapter 1 raises the pattern here and returns to it with fuller strategies in Chapter 9.

(a) The association between academic effort and group identity in some peer environments;

(b) the immediate social cost of visible effort set against the delayed reward of academic success;

(c) classroom behaviors, including silence and incomplete work, that often disguise fear of peer judgment rather than a lack of ability.

V. Growth Mindset as a Counter-Narrative

A fixed view of intelligence treats ability as a trait a person either has or lacks, and that view leaves no room for growth after a wrong answer. Effort, strategy, and practice changed measured ability over time among students who received growth-mindset instruction paired with real skill-building support (Dweck, 2006). Neural connections form in response to struggle rather than in its absence, a claim grounded in brain research on plasticity (Boaler, 2016). A growth mindset does not promise instant success; it promises that current struggle carries information about where to go next in practice.

(a) The difference between a fixed mindset ("I can't do math") and a growth mindset ("I can't do this yet");

(b) brain plasticity and the formation of new neural pathways through practice (Boaler, 2016);

(c) the role of language, including teacher and parent phrasing, in reinforcing one mindset over the other (Dweck, 2006).

VI. Struggle Is Not a Verdict: Redefining Intelligence

Intelligence, in the framework this chapter builds, describes a capacity that grows rather than a fixed score a student carries from birth. A wrong answer indicates a gap in strategy or vocabulary; it does not prove low ability. Students who understand this distinction persist longer on hard problems and recover faster from setbacks, because the setback no longer threatens their identity. Chapter 2 picks up this thread and builds a fuller mindset toolkit around it.

(a) The reframing of struggle as data rather than judgment;

(b) the separation of self-worth from a single test score or homework grade;

(c) a closing challenge for students: rewrite one personal "I'm not smart" statement into a growth-oriented alternative.

Research Foundations Cited in This Chapter

Ashcraft, M. H. (2002). Math anxiety: Personal, educational, and cognitive consequences. *Current Directions in Psychological Science*, *11*(5), 181–185.

Boaler, J. (2016). *Mathematical mindsets: Unleashing students' potential through creative math, inspiring messages, and innovative teaching*. Jossey-Bass.

Dweck, C. S. (2006). *Mindset: The new psychology of success*. Random House.

Steele, C. M., & Aronson, J. (1995). Stereotype threat and the intellectual test performance of African Americans. *Journal of Personality and Social Psychology*, *69*(5), 797–811.

Supplemental Features for This Chapter

Math Mindset Moment: a short boxed feature closes the chapter with a single reframe students can carry into their next math class: "I am not bad at math. I am early in my practice." The feature pairs the phrase with a two-minute breathing or grounding exercise students can use before a quiz or cold-call question.

Student Success Story: a brief profile introduces a composite student, built from common classroom patterns rather than from a single identifiable person, who moved from hiding answers to raising a hand within a single semester. The story names the specific fear the student faced and the specific classroom practice that interrupted it, giving readers a concrete model rather than an abstract promise.

Reflection Questions:

(a) describe a moment when you avoided a math question in class because you feared being wrong;

(b) name one thing you have said about math that protected you from looking like you were trying;

(c) identify a time a mistake in math taught you something a correct answer could not have taught you;

(d) explain, in one paragraph, the difference between "I am bad at math" and "I have not practiced this skill yet";

(e) identify one classmate, teacher, or family member you could ask for support the next time math feels overwhelming.

Chapter Takeaway

Fear explains far more classroom silence than an actual dislike of mathematics ever could. Students protect themselves with indifference, avoidance, and silence because those responses feel safer than a visible attempt that might fail. Growth mindset research, stereotype threat research, and identity research each point to the same conclusion: the belief a student holds about their own capacity shapes the effort that student is willing to spend (Dweck, 2006; Steele & Aronson, 1995). Struggle, in this framework, becomes evidence of learning in progress rather than evidence of low intelligence. Chapter 2 builds on this foundation and turns the cycle named here into a set of tools for rewriting the narrative a student tells about themselves.

Chapter 2: Rewriting the Narrative

Changing the Way Students Think About Math

Getting Started

Everyone has a story they tell themselves. Some students say, “I’m just not a math person,” while others believe that only the smartest students can excel in mathematics. Stories like these become powerful because they shape expectations and influence actions. Fortunately, every story can be rewritten. Your past experiences do not determine your future success. A disappointing report card, a failed quiz, or a difficult teacher does not define your mathematical ability. What matters most is the decision to believe that growth is possible.

Think about someone learning to play the piano. No one expects to perform a difficult song perfectly on the first attempt. Mistakes are expected because learning requires practice. Mathematics works the same way. Consider a student learning to calculate interest on a savings account. At first, the formulas may seem confusing, but after repeated practice, the calculations become easier. Before long, that same student can determine how much money an investment will grow over time. Confidence is built one successful experience after another.

As you rewrite your personal narrative, replace limiting beliefs with empowering ones. Instead of saying, “I can’t do this,” begin saying, “I can’t do this yet.” Every problem solved strengthens your mathematical muscles. Your story is still being written, and every chapter offers an opportunity to become stronger, wiser, and more confident than before.

Real-Life Math Connection

Imagine you deposit $500 into a savings account earning 5% annual interest.

Solution

$I = P \times R \times T$

$I = 500(.05)(3)$

$I = \$75$

After three years:

$\$500 + \$75 = \$575$

Learning one formula can help you understand how money grows over time.

Chapter Purpose and Guiding Question

How can a student replace a fixed narrative about mathematical ability with one rooted in growth?

Chapter 2 turns from naming fear to rewriting the story a student tells about their own ability in mathematics. A narrative, once repeated enough times, starts to feel like a fact rather than a choice, and many students carry a narrative built from someone else's low expectations. Cultural messages, classroom experiences, and a string of low grades combine to write that narrative long before a student develops the tools to question it. Implicit theories of intelligence shaped the goals students pursued and the strategies they chose when problems became difficult, with some students working to prove their ability and others to build it (Dweck & Leggett, 1988). A single failed quiz seldom causes lasting damage on its own; repetition does the damage, turning one bad day into a settled identity. Structural brain scans showed that a region associated with spatial learning grew larger with years of practice among London taxi drivers, providing evidence that the brain continues to change well into adulthood (Maguire et al., 2000). Chapter 2 uses that evidence to give students a replacement narrative, one built on the starting point as a beginning rather than a verdict. Chapter 1 named the fear cycle; Chapter 2 supplies the story that breaks it, and closes with a personal rewriting exercise that carries into every chapter that follows.

Learning Objectives

Chapter 2 sets five learning outcomes for students to reach before its final page:

(a) trace one personal math narrative back to a specific classroom or cultural origin;

(b) distinguish between a fixed mindset goal ("prove I'm smart") and a growth mindset goal ("get smarter") (Dweck & Leggett, 1988);

(c) explain, using brain research, why practice changes the brain at any age (Maguire et al., 2000);

(d) define productive struggle and describe why it differs from unproductive frustration (National Council of Teachers of Mathematics [NCTM], 2014);

(e) rewrite one fixed-mindset statement about their own math ability into a growth-oriented alternative.

Section Outline

I. Where the Narrative Comes From: Cultural and Classroom Origins

A student's belief about math seldom originates inside a math problem; it originates inside a room, a comment, or a comparison. Cultural messages about who belongs in advanced math classes travel through family conversations, media representation, and the unspoken composition of gifted and AP rosters. A single offhand comment from a teacher, spoken once and never thought of again, can anchor a student's sense of ability for years. Stereotype threat research demonstrated that performance dropped when a test was framed as diagnostic of intellectual ability among students aware of a negative stereotype about their group (Steele & Aronson, 1995).

(a) Family and community messages about who is or is not a "math person";

(b) classroom seating, grouping, and tracking decisions that signal ability before a student has proven anything;

(c) stereotype threat and its measurable effect on test performance under identity-diagnostic conditions (Steele & Aronson, 1995).

II. Repeated Failure and the Fixed-Mindset Trap

One low grade seldom convinces a student they lack ability; a pattern of low grades does. Students oriented toward proving their intelligence read each new failure as additional evidence for a verdict they have already reached about themselves, while students oriented toward building their intelligence read the same failure as a signal to adjust strategy (Dweck & Leggett, 1988). The trap closes when a student stops attempting hard problems altogether, since an unattempted problem cannot supply new evidence against a fragile identity. Avoidance then masquerades as disinterest, protecting the student from a risk that feels larger than it is.

(a) The shift from "I failed this test" to "I am a failure," a shift that compounds over repeated setbacks;

(b) goal orientation: proving intelligence versus building it (Dweck & Leggett, 1988);

(c) avoidance as self-protection, including skipped homework and abandoned problems.

III. The Neuroscience of Growth: How the Brain Builds Math Ability

The brain does not finish developing in childhood and then stop; it continues to reorganize itself in response to sustained practice at any age. Structural scans of licensed taxi drivers showed measurable growth in a brain region tied to spatial navigation after years of route practice, growth that tracked with time on the job

(Maguire et al., 2000). Mathematical practice recruits its own networks in a comparable way, strengthening pathways each time a student works through a problem instead of avoiding it (Boaler, 2016). A struggling student is not a fixed case; a struggling student is a brain mid-construction.

(a) Neuroplasticity as a lifelong process rather than a window that closes after childhood;

(b) the taxi driver hippocampus study as concrete evidence of structural brain change through practice (Maguire et al., 2000);

(c) the application of plasticity research to mathematics practice and skill-building (Boaler, 2016).

IV. Productive Struggle: Redefining What Effort Looks Like

Productive struggle asks a student to sit inside confusion long enough to work through it, an experience many classrooms train students to avoid rather than tolerate. Task engagement that emphasizes reasoning and relationships among ideas produced deeper mathematical understanding than task engagement built around speed and correct answers alone (National Council of Teachers of Mathematics [NCTM], 2014). A teacher who rescues a student from every difficult moment removes the very friction that would have led to new understanding. Structure is what turns struggle productive: a clear task, a reasonable amount of time, and a teacher who asks questions instead of supplying answers.

(a) The difference between productive struggle and unproductive frustration;

(b) teacher moves that support struggle, including wait time and scaffolded questioning (NCTM, 2014);

(c) confusion and error treated as expected steps in learning rather than signs of failure.

V. Two Mindsets, Two Classrooms: Fixed vs. Growth Language

Language shapes a mindset before a student notices the shift. Praise directed at a fixed trait ("you're so smart") taught students to protect that label, while praise directed at process ("you worked through that with real effort") taught students to keep working (Dweck, 2006). A classroom that celebrates correct answers alone teaches the same lesson as a classroom that punishes wrong ones: safety lives in silence. A teacher's or parent's small phrase changes, repeated across a semester, accumulate into a student's internal voice.

(a) Trait praise versus process praise and their long-term effects on effort (Dweck, 2006);

(b) classroom norms that reward correct answers alone versus norms that reward reasoning and revision;

(c) the accumulation of small language choices into a student's internal self-talk over time.

VI. Writing a New Personal Narrative

A student cannot erase a difficult math history, and this chapter does not ask them to try. A student can, with structure and repetition, write a new chapter onto that history starting today. The rewriting exercise below asks students to name their old narrative, name the evidence that built it, and draft a new sentence grounded in the neuroscience and mindset research covered in this chapter. Chapter 3 builds on this new narrative, giving students the vocabulary they need to talk about math with precision and confidence.

(a) Naming the old narrative in a single sentence;

(b) tracing that narrative back to its origin: a comment, a grade, or a comparison;

(c) drafting a replacement sentence grounded in evidence rather than emotion alone.

Research Foundations Cited in This Chapter

Boaler, J. (2016). *Mathematical mindsets: Unleashing students' potential through creative math, inspiring messages, and innovative teaching*. Jossey-Bass.

Dweck, C. S. (2006). *Mindset: The new psychology of success*. Random House.

Dweck, C. S., & Leggett, E. L. (1988). A social-cognitive approach to motivation and personality. *Psychological Review*, *95*(2), 256–273.

Maguire, E. A., Gadian, D. G., Johnsrude, I. S., Good, C. D., Ashburner, J., Frackowiak, R. S., & Frith, C. D. (2000). Navigation-related structural change in the hippocampi of taxi drivers. *Proceedings of the National Academy of Sciences*, *97*(8), 4398–4403.

National Council of Teachers of Mathematics. (2014). *Principles to actions: Ensuring mathematical success for all*. NCTM.

Steele, C. M., & Aronson, J. (1995). Stereotype threat and the intellectual test performance of African Americans. *Journal of Personality and Social Psychology*, *69*(5), 797–811.

Supplemental Features for This Chapter

Math Mindset Moment: a short boxed feature gives students a two-line reframe to say after a mistake: "My brain just grew. That is what this feeling means." The feature links the phrase to the brain-based evidence covered in Section III, so the affirmation carries scientific weight rather than empty encouragement.

Student Success Story: a composite profile follows a student who spent a semester avoiding word problems out of fear of being wrong, then used the personal-narrative rewrite from Section VI to change how they responded to a hard question. The story shows the specific sentence the student replaced and the new sentence that replaced it, giving readers language they can use in their own practice.

Reflection Questions:

(a) write the math narrative you have carried since elementary school in one or two sentences;

(b) identify the specific comment, grade, or comparison that helped build that narrative;

(c) describe a moment this year when a problem felt like productive struggle rather than pointless frustration;

(d) explain, in your own words, what happens inside the brain when you practice a hard skill (Maguire et al., 2000);

(e) draft a new sentence about your math ability, grounded in evidence rather than an old grade or comment.

Chapter Takeaway

A narrative built from cultural messages, classroom experiences, and repeated failure feels permanent long before a student ever tests whether it is true. Brain research and mindset research each supplied evidence against that permanence, showing a capacity that grows through structured practice rather than a trait fixed at birth (Dweck & Leggett, 1988; Maguire et al., 2000). Productive struggle gave that evidence a daily practice, turning confusion into the exact condition under which new understanding forms (NCTM, 2014). A rewritten narrative does not remove every hard math problem from a student's path; it changes what that student believes a hard problem means. Chapter 3 turns from mindset to language, teaching students that mathematical vocabulary is the next tool for turning belief into visible skill.

Chapter 3: Math Has a Language

Why Vocabulary Matters in Mathematics

Getting Started

Every subject has its own language. Science has scientific vocabulary, music has notes and symbols, and mathematics has numbers, expressions, equations, variables, and formulas. Many students struggle not because they lack intelligence but because they have not yet learned to speak the language of mathematics. Once students understand the vocabulary, symbols, and relationships between concepts, solving problems becomes much more manageable.

Imagine traveling to another country where everyone speaks a different language. At first, communication feels impossible. However, after learning a few words each day, conversations become easier. Mathematics works the same way. Learning words such as “coefficient,” “ratio,” “slope,” “probability,” and “proportion” gives students the tools they need to understand increasingly advanced ideas. A weather forecast predicting a 70% chance of rain becomes easier to act on once you understand probability, helping you decide whether to carry an umbrella or change outdoor plans.

Learning the language of mathematics takes patience and consistency. Read mathematical problems carefully, underline important vocabulary, and ask questions whenever something is unclear. The more fluent you become in the language of mathematics, the more confident you will feel when facing challenging problems. Mathematics is no longer a mystery once you understand how to speak its language.

Real-Life Math Connection

Out of the last 20 days with cloud cover like today's, rain fell on 14 of them. Based on that history, what is the probability that it rains today?

Solution

Probability = Favorable Outcomes ÷ Total Outcomes

P(rain) = 14 ÷ 20

= 0.70

= 70%

Fourteen days of weather history turn into a single number you can plan your umbrella around.

Chapter Purpose and Guiding Question

In what ways does mathematical vocabulary function as a language a student must learn before mathematics itself makes sense?

Chapter 3 turns from belief to language, the next barrier standing between a student and a math problem that finally makes sense. Mathematics carries a vocabulary load most students never see named, packed with words that mean one thing in daily conversation and another in a math class. A student can understand the underlying concept and still fail a problem because a single word blocked the entrance to that understanding. Vocabulary functioned as one part of a wider language demand tied to mathematics achievement, sitting alongside symbols, syntax, and reading comprehension (Thompson & Rubenstein, 2000). Words such as evaluate, factor, quotient, variable, and slope carry precise technical definitions that a student must learn the way a foreign-language student learns a new word, not the way a native speaker absorbs a familiar one. Reading a math problem demanded a distinct set of skills from reading a story, since numerals, symbols, and words combined to carry meaning that any one of the three could not carry alone (Adams, 2003). Chapter 3 treats vocabulary instruction as a foundation skill rather than an afterthought squeezed in before a test. Strong vocabulary instruction gives students a stable base under the mindset work of Chapters 1 and 2, since confidence has little room to grow around words a student cannot decode. The remaining sections provide a comprehensive toolkit for teaching, practicing, and reinforcing math vocabulary throughout the school year.

Learning Objectives

Chapter 3 sets five learning outcomes for students to reach before its final page:

(a) Define at least five core academic math terms introduced in the current unit, including evaluate, factor, quotient, variable, and slope;

(b) identify math words that carry a different meaning in everyday conversation than they carry in a math class;

(c) build one Frayer model or graphic organizer for a new math term (Frayer et al., 1969);

(d) maintain a vocabulary journal entry for each new unit, including a definition, an example, and a visual (Marzano & Pickering, 2005);

(e) use a classroom word wall or personal reference tool to support vocabulary recall during independent work.

Section Outline

I. Math as a Second Language: Why Vocabulary Blocks Understanding

Mathematics functions as a second language for most students, complete with its own grammar, symbols, and vocabulary that a spoken-language classroom would never assume students already know. A student who cannot translate the word quotient into an operation cannot access a problem, regardless of how strong that student's number sense might be. Vocabulary gaps produced errors that appeared to be conceptual gaps on the surface, sending some students into remediation for a skill they had once possessed once the word blocking them was named and taught (Thompson & Rubenstein, 2000). Treating vocabulary as a taught skill, not an assumed one, closes that gap before it produces a false diagnosis of a student's mathematical ability.

(a) Vocabulary as a prerequisite skill rather than a byproduct of problem-solving;

(b) the risk of misdiagnosing a vocabulary gap as a conceptual gap (Thompson & Rubenstein, 2000);

(c) explicit, direct instruction of new terms before a unit begins rather than during a test.

II. Multiple Meanings and the Trap of Everyday Words

Some of the most damaging math vocabulary is not new at all; it already lives in a student's everyday vocabulary with a competing meaning. The word table means a piece of furniture in one context and an organized display of data in another, and a student's brain must learn to switch between the two on command. Words including product, difference, even, odd, and rational carry a technical meaning that can contradict the word's common meaning, creating confusion that has nothing to do with a student's math skill. Reading comprehension research identified this category of words as a distinct and predictable source of error across grade levels, separate from unfamiliar technical terms (Adams, 2003).

(a) Homograph math terms with competing everyday meanings, such as table, product, and even;

(b) the classroom practice of naming the everyday meaning first, then the math meaning, to make the switch explicit;

(c) reading comprehension breakdowns traced to multiple-meaning words rather than unfamiliar technical vocabulary (Adams, 2003).

III. Building Academic Math Vocabulary as a System

Vocabulary instruction works best as a planned system rather than a collection of one-time definitions written on a whiteboard and erased before the next class period. A six-step process for teaching new terms, moving from a plain-language explanation through student restatement, non-linguistic representation, and repeated review, built durable academic vocabulary across subject areas, including mathematics (Marzano & Pickering, 2005). A term taught once and never revisited fades from memory at nearly the same rate, whether it came from a math class or any other subject. Systematic review, spaced across days and weeks rather than crammed before a single test, keeps academic math vocabulary available for use during problem-solving, rather than locked in short-term memory.

(a) A plain-language explanation offered before the formal textbook definition;

(b) student restatement of a term in their own words as a check for understanding;

(c) non-linguistic representation, including sketches, diagrams, or gestures tied to a term;

(d) spaced review that revisits vocabulary across multiple weeks rather than a single lesson (Marzano & Pickering, 2005).

IV. The Frayer Model and Graphic Organizers

The Frayer model organizes a new term into four quadrants: a definition, essential characteristics, examples, and non-examples, giving students a consistent structure for building a complete concept rather than a single sentence. Concept mastery improved when students sorted examples from non-examples themselves instead of copying a definition from a textbook (Frayer et al., 1969). A completed Frayer model for the term variable might list "a letter representing an unknown number" as the definition, list "can change value" as a characteristic, list 3x and y + 2 as examples, and list the number 7 alone as a non-example. Graphic organizers built on this structure give students a repeatable process they can apply to any new term without waiting for a teacher to model it each time.

(a) The four Frayer model quadrants: definition, characteristics, examples, and non-examples (Frayer et al., 1969);

(b) a worked example for the term variable, presenting a definition, a characteristic, sample expressions, and a non-example together;

(c) non-example selection, which forces a student to articulate the boundary of a concept rather than its center alone.

V. Vocabulary Journals and Visual Strategies

A vocabulary journal provides a student with a personal, cumulative record of every math term learned over a school year, turning scattered definitions into a single searchable reference. Each entry pairs a formal definition with a student-written explanation and a visual cue, following the same principle of non-linguistic representation shown to strengthen retention (Marzano & Pickering, 2005). Color-coding terms by unit or operation type adds a visual layer that helps some students recall terms more quickly during a timed assessment. A journal maintained across an entire year becomes a personal reference text a student can consult on their own, reducing the moments when a forgotten word stops a problem cold.

(a) A consistent entry format: term, definition, student explanation, and visual cue;

(b) color-coding according to unit, operation, or concept family;

(c) a cumulative structure that carries forward across the full school year rather than resetting each unit.

VI. Word Walls and Reference Tools as Daily Support

A word wall keeps active vocabulary visible during the exact moments a student needs it most: mid-problem, mid-sentence, mid-explanation. Displayed terms paired with a definition, an example, and a visual give a student a resource to consult without raising a hand or interrupting a lesson to ask what a word means. Reference tools extend that same support into homework and independent practice, where a teacher is not present to answer a quick vocabulary question. A classroom that treats its word wall as a living document, revised unit after unit, keeps the tool relevant instead of turning it into background decoration.

(a) A word wall organized according to unit, updated as new terms are introduced and retired;

(b) a personal reference sheet or bookmark students can carry between class and home;

(c) QR codes or short links connected to a term, offering a quick video reminder for terms that resist a single clear definition.

Research Foundations Cited in This Chapter

Adams, T. L. (2003). Reading mathematics: More than words can say. *The Reading Teacher*, *56*(8), 786–795.

Frayer, D. A., Frederick, W. C., & Klausmeier, H. J. (1969). *A schema for testing the level of concept mastery* (Technical Report No. 16). University of Wisconsin, Research and Development Center for Cognitive Learning.

Marzano, R. J., & Pickering, D. J. (2005). *Building academic vocabulary: Teacher's manual.* Association for Supervision and Curriculum Development.

Thompson, D. R., & Rubenstein, R. N. (2000). Learning mathematics vocabulary: Potential pitfalls and instructional strategies. *Mathematics Teacher*, *93*(7), 568–574.

Supplemental Features for This Chapter

Math Mindset Moment: a short boxed feature challenges students to use one new vocabulary word correctly in a sentence about their own life before using it in a math problem, turning an abstract term into personal language. The feature closes with the line: "I do not just solve math. I speak it."

Student Success Story: a composite profile follows a student who confused quotient and product on quiz after quiz until a Frayer model exercise separated the two terms into distinct visual entries. The story traces the specific moment the confusion cleared and the specific study habit, a five-minute journal review each day, that kept the distinction in place through the rest of the year.

Reflection Questions:

(a) list three math words that confused you this year and explain why each one caused trouble;

(b) identify one everyday word that means something different in math class, such as table or product;

(c) build a Frayer model for one term from your current unit, including an example and a non-example;

(d) describe how a vocabulary journal or word wall could have helped you avoid a specific past mistake;

(e) teach one math term to a family member using a definition, an example, and a picture.

Chapter Takeaway

Mathematics carries a vocabulary load that many classrooms leave untaught, treating technical terms as though students will absorb them through exposure alone. Vocabulary functions as a distinct skill, one that responds to direct instruction, visual representation, and spaced review with the same reliability as any other academic skill (Marzano & Pickering, 2005; Thompson & Rubenstein, 2000). Multiple-meaning words, Frayer models, vocabulary journals, and word walls each address the same barrier from a different angle, providing teachers with multiple entry points for the same goal (Adams, 2003; Frayer et al., 1969). A student who can name a concept with precision gains a tool for thinking, beyond a term memorized for a single quiz. Chapter 4 turns from vocabulary to organization, showing students how to structure their notes and materials so the language students just learned stays accessible long after a unit ends.

Chapter 4: Organization Is a Math Superpower

Building Systems for Success

Getting Started

Success in mathematics is often determined long before students begin solving problems, beginning with organization. Organized students know where to find their notes, complete assignments on time, keep track of formulas, and develop study routines that reduce stress before quizzes and tests. Organization is not about being perfect; it is about creating systems that let your brain focus on solving problems rather than searching for missing papers or forgotten homework. Just as an engineer carefully organizes blueprints before constructing a building, successful math students prepare themselves before tackling difficult concepts.

Organization also plays a major role in everyday life. Imagine planning a family vacation. You must calculate travel time, estimate gas expenses, budget for meals, determine hotel costs, and organize daily activities. Without a plan, expenses quickly exceed the budget. Similarly, a student preparing for a math test who organizes formulas, vocabulary, practice problems, and study time is far more likely to succeed than one who studies the night before. Whether you're balancing a monthly budget or organizing your school assignments, mathematics rewards preparation and planning.

Being organized builds confidence by eliminating unnecessary distractions. Create folders for each unit, keep a dedicated math notebook, write assignments in a planner, and review your notes for a few minutes each evening. Small habits practiced consistently produce remarkable results. Organization transforms mathematics from feeling overwhelming into becoming manageable, one step at a time.

Real-Life Math Connection

Imagine your family has a monthly budget.

Solution

Income:

$4,800

Expenses

Rent: $1,500

Food: $700

Transportation: $450

Utilities: $350

Savings: $500

Entertainment: $300

Total Expenses

= $3,800

Money Remaining

$4,800 − $3,800 = $1,000

Organization helps families make informed financial decisions.

Chapter Purpose and Guiding Question

How does organization function as a cognitive skill that supports mathematical achievement?

Chapter 4 treats organization as a cognitive skill worth direct instruction, not a personality trait some students have, and others do not. A messy binder seldom signals laziness; it often signals an executive function skill that has never been explicitly taught. Executive function measures, including inhibition, working memory, and cognitive flexibility, accounted for a meaningful share of the variance in children's mathematics ability across multiple tasks (Bull & Scerif, 2001). Organization draws on the same set of skills: a student must hold a plan in mind, resist the pull toward a faster shortcut, and switch between steps in a fixed sequence without losing the thread. A student who cannot find last week's notes loses more than time; they also lose the visible record of their own progress. Executive skills, including planning, organization, and time management, develop on a predictable yet uneven timeline throughout childhood and adolescence, which means some students need direct instruction long after their peers have absorbed the system without it (Dawson & Guare, 2010). Chapter 4 gives teachers and students a concrete system: color-coded binders, structured notes, and portfolios designed to make progress visible rather than buried in a backpack. A student who can see their own growth across a semester carries evidence against the fear named in Chapter 1. The remaining sections build that system one piece at a time, starting with the notebook a student opens every day.

Learning Objectives

Chapter 4 sets five learning outcomes for students to reach before its final page:

(a) Explain why organization functions as an executive function skill rather than a fixed personality trait (Dawson & Guare, 2010);

(b) set up a color-coded math binder with sections for notes, practice, assessments, and reference tools;

(c) complete a daily note-taking header including name, date, objective, and math concept;

(d) build one interactive notebook page or graphic organizer for a current unit;

(e) track a full, multi-step solution in writing, showing every step rather than just the final answer.

Section Outline

I. Organization as an Executive Function Skill

Organization belongs to a family of mental skills called executive functions, the same skills responsible for planning, working memory, and self-control. Executive functions make it possible to hold a goal in mind, block out a distraction, and adjust a plan when the first approach fails, a description that matches what a student does every time they solve a multi-step math problem (Diamond, 2013). A student with weak organizational habits often has strong mathematical reasoning trapped behind a system that keeps losing the evidence of that reasoning. Naming organization as a teachable skill, rather than a fixed trait, opens the door to direct instruction rather than lowered expectations.

(a) Planning, working memory, and cognitive flexibility as the core executive functions tied to math performance (Bull & Scerif, 2001; Diamond, 2013);

(b) the difference between a motivation problem and an unmet organizational-skill need;

(c) direct instruction in organization as a classroom priority, not an assumed background skill.

II. Daily Note-Taking Expectations: Name, Date, Objective, Concept

A consistent notebook header, written the same way every day, does more work than it appears to do on the surface. Name, date, objective, and math concept give every page a searchable identity, turning a stack of loose paper into a navigable record a student can use months later. Review behaviors, including how notes are organized and revisited, carry as much weight for later recall as the note-taking itself, which means the header a student writes today shapes how useful that page will be in May (Kiewra, 1989). A missing header costs nothing on the day it is written, but a great deal the day a student needs to find that page again.

(a) A fixed header format used across every subject and every day: name, date, objective, concept;

(b) the objective written in student language, not copied word for word from a standard;

(c) a consistent placement for the header so a student can locate it without searching the full page.

III. The Color-Coded Math Binder: A System, Not a Decoration

A color-coded binder works because color provides the brain with a fast retrieval cue that text alone cannot provide. Each unit or strand is assigned a fixed color, carried across notes, handouts, and practice problems, so a student can locate an algebra section by its color before reading a single word. A binder built this way turns a five-minute search for last Tuesday's notes into a five-second flip to the correct tab. The system fails the moment it becomes decoration rather than function, so teachers do best when they model the system every day rather than assigning it once at the start of a term.

(a) One fixed color per unit or strand, applied to tabs, notes, and handouts alike;

(b) a consistent section order: notes, classwork, homework, assessments, reference tools;

(c) teacher modeling of the system across the term, not a single September lesson.

IV. Graphic Organizers and Interactive Notebooks

An interactive notebook turns passive storage into active practice, asking a student to fold, sort, glue, and annotate material instead of leaving it filed away untouched. Graphic organizers give shape to a concept before a student memorizes a single formula, mapping relationships between terms, steps, and examples in a format the brain can scan at a glance. A left-side, right-side notebook structure, with input on one page and processing or reflection on the facing page, gives every new concept a built-in space for a student's own thinking. The physical act of building an organizer, rather than receiving a finished one, does part of the learning work before a single practice problem begins.

(a) Interactive notebook structures with input on one side and processing or reflection on the other;

(b) graphic organizers that map relationships between terms, steps, and worked examples;

(c) student-built organizers rather than teacher-completed handouts, so the construction itself becomes practice.

V. Tracking Every Step: Making Thinking Visible

A final answer shows what a student concluded; a tracked solution shows how a student got there, and the second record carries far more diagnostic value than the first. Writing out every step, including steps a student can complete in their head, builds a habit of external memory that frees working memory for the harder parts of a problem (Diamond, 2013). A teacher reading a solution with every step shown can locate the exact step where an error entered, turning a wrong answer into a specific, teachable moment instead of a vague signal that something went wrong. Step tracking also protects a student during a test, since partial credit and self-correction both depend on a visible trail of reasoning.

(a) Writing every step, including steps that feel too simple to record;

(b) using tracked steps to locate the exact point where an error entered a solution;

(c) partial credit and self-correction as direct payoffs of a visible solution trail.

VI. Building the Math Portfolio: Organization Over Time

A math portfolio collects evidence of growth across a semester or a year, turning a series of disconnected assignments into a single visible arc of progress. Selected work samples, paired with a short student reflection, show a skill developing over time in a way a single grade never can. A portfolio built over a full term provides a student with concrete evidence to counter the old narrative outlined in Chapter 2, since evidence of growth is harder to dismiss than a general feeling of improvement. Chapter 5 builds on this system, teaching students how to turn organized materials into notes that build understanding rather than record it after the fact.

(a) Selected work samples chosen across a semester rather than gathered at the very end;

(b) a short student reflection attached to each portfolio entry;

(c) a visible arc of growth that a student can present as evidence during a parent conference or self-assessment.

Research Foundations Cited in This Chapter

Bull, R., & Scerif, G. (2001). Executive functioning as a predictor of children's mathematics ability: Inhibition, switching, and working memory. *Developmental Neuropsychology*, *19*(3), 273–293.

Dawson, P., & Guare, R. (2010). *Executive skills in children and adolescents: A practical guide to assessment and intervention* (2nd ed.). The Guilford Press.

Diamond, A. (2013). Executive functions. *Annual Review of Psychology*, *64*, 135–168.

Kiewra, K. A. (1989). A review of note taking: The encoding-storage paradigm and beyond. *Educational Psychology Review*, *1*(2), 147–172.

Supplemental Features for This Chapter

Math Mindset Moment: a short, boxed feature asks students to color-code one page of notes before the next class and, in one sentence, note whether the page became easier to use. The feature closes with the line: "A system is not extra work. A system is work I do once, not work I repeat forever."

Student Success Story: A composite profile follows a student who lost points across a full semester for missing homework that had, in fact, been completed and buried in a backpack. The story traces the specific week a color-coded binder and a fixed notebook header ended the pattern, showing the grade change that followed within a single grading period.

Reflection Questions:

(a) Describe your current system, if you have one, for organizing math notes and assignments;

(b) identify one assignment you lost points on this year because of a missing or misplaced paper rather than a math error;

(c) choose one color you will use for your current math unit and explain your reason for choosing it;

(d) explain why writing every step of a solution might help a teacher more than writing the final answer alone;

(e) describe one piece of work you would choose for a math portfolio and explain what it shows about your growth.

Chapter Takeaway

Organization functions as a learnable executive skill, not a fixed trait some students carry, and others lack (Dawson & Guare, 2010; Diamond, 2013). A consistent header, a color-coded binder, an interactive notebook, and a tracked solution each remove a small barrier between a student's thinking and the record of that thinking. Removing enough small barriers changes what a student's effort produces, turning scattered work into a visible, buildable record of growth. A portfolio built across a term gives that growth a physical form a student can hold, point to, and defend. Chapter 5 turns from the structure around a student's notes to the notes themselves, teaching the specific writing habits that turn a math notebook into an active learning tool.

Chapter 5: The Power of Note-Taking

Learning Mathematics Through Writing

Getting Started

Many students believe that taking notes simply means copying what the teacher writes on the board. Effective note taking is much more than that: an active learning strategy that helps organize information, identify important concepts, and strengthen memory. Writing mathematical procedures in your own words forces your brain to process the information more deeply, making it easier to recall during homework and tests. Your notebook becomes a personalized textbook filled with examples, reminders, and strategies that make learning easier.

Consider someone following a recipe for baking a cake. If they skip writing down ingredient measurements or directions, they are likely to forget an important step. Mathematics works the same way. Imagine learning how to solve proportions while calculating ingredients for a recipe that serves twelve people instead of six. Careful notes showing each step allow you to adjust ingredient amounts accurately. Likewise, when calculating discounts during holiday shopping or comparing prices at different stores, well-organized notes on percentages can help you make smart financial decisions.

Develop the habit of writing key vocabulary, formulas, worked examples, and questions that arise during class. Use colors to highlight important ideas and leave space for additional explanations later. Reviewing your notes for just ten minutes each day strengthens understanding far more than trying to relearn everything the night before an exam. Great note taking transforms confusion into clarity and builds confidence with every lesson.

Real-Life Math Connection

A recipe serves 6 people, but you're cooking for 15.

Solution

Multiplier

$15 \div 6 = 2.5$

If the recipe calls for:

2 cups of flour

$2 \times 2.5 = 5$ cups

Your notes on proportions make cooking easier.

Suggested Math Concepts

- Ratios
- Proportions
- Multiplication
- Unit rates

Chapter Purpose and Guiding Question

How does active note taking transform a math notebook into a tool for understanding rather than a record of copied work?

Chapter 5 turns organized materials into an active tool for learning, teaching students to write their way into mathematical understanding rather than copy their way through a lesson. Note taking, treated as a passive transcription task, produces a page a student seldom opens again; note taking, treated as an act of thinking, produces a page a student returns to before every quiz. Review behaviors built on well-organized notes supported long-term recall to the same degree as the original note-taking act, which means the structure of a page matters as much as the decision to write on it (Kiewra, 1989). Strong math students write more than weak math students do during the same lesson, not because they understand more from the start but because writing itself builds the understanding that shows up later. A student who annotates an example, tracks every step, and reflects on an error in writing processes that problem several times over, while a student who copies a final answer processes it once. Mathematics classrooms have not always treated writing as central to the subject, even though writing forced a level of clarity that talking alone seldom produced, since a written sentence cannot hide behind a vague gesture toward the board (Countryman, 1992). Chapter 5 gives students five specific writing habits: Cornell notes adapted for math, annotation, full work shown, error reflections, and a running math journal. Each habit turns a math notebook from a record of what happened into evidence of how a student thinks. The remaining sections teach each habit in turn, building on the organizational systems introduced in Chapter 4.

Learning Objectives

Chapter 5 sets five learning outcomes for students to reach before its final page:

(a) set up a two-column Cornell-style page for math notes, including a cue column and a summary section (Pauk, 1962);

(b) annotate one worked example, marking each step with a short note explaining the reasoning behind it;

(c) show complete work for every problem on an assignment, rather than a final answer alone;

(d) write a short reflection explaining the cause of one mistake from a recent assignment or test;

(e) maintain a math journal entry connecting a new concept to a real example or a personal observation (Countryman, 1992).

Section Outline

I. Note-Taking as a Thinking Tool, Not a Copying Task

Copying a teacher's board work word for word keeps a student's hand busy while leaving the brain idle for the most part, since transcription demands attention but not analysis. Generative note taking, the kind that asks a student to summarize, question, or restate material in their own words, produced stronger recall than verbatim copying across a wide range of subjects, including mathematics (Kiewra, 1989). A student who pauses to write "why does this step work" beside a line of algebra engages a different mental process than a student who reproduces the line without a pause. Chapter 5 asks students to treat every page of notes as a conversation with the material, not a transcript of what a teacher said.

(a) The difference between verbatim copying and generative note taking (Kiewra, 1989);

(b) short margin questions that turn a passive line of notes into an active one;

(c) treating a notes page as a record of thinking, not only a record of content.

II. The Cornell Method Adapted for Mathematics

The Cornell system divides a page into a narrow cue column, a wide note-taking column, and a summary section at the bottom, a structure built to separate the moment of capture from the moment of review (Pauk, 1962). In a math adaptation, the note-taking column holds the worked example one line at a time, the cue column holds a short label for each step, such as "distribute" or "combine like terms," and the summary section holds a one- or two-sentence restatement of the method in the student's own words. Covering the note-taking column and quizzing from the cue column turns a static page into a retrieval practice tool a student can use the night before a test. A blank summary section signals unfinished thinking, which gives a teacher a quick visual check of which students need a second look at a lesson.

(a) A note-taking column for the full worked example, written one line at a time;

(b) a cue column with short labels naming each step or strategy;

(c) a summary section written in the student's own words, completed before the notes are considered finished (Pauk, 1962).

III. Annotating Worked Examples

Annotation asks a student to mark a worked example, writing on or beside it to capture the reasoning that a textbook or teacher often leaves unspoken between one line and the next. A circled operation, a bracket around a substituted value, or a short note reading "why negative" turns a finished example into a map of decisions rather than a sequence of results. Annotated examples give a student a model to imitate during independent practice, since the reasoning that produced each step sits visible on the page instead of hidden inside a teacher's head. A student building this habit slows down during instruction, and that slower pace pays for itself the first time an assignment asks for a similar problem without a model in sight.

(a) Circling or boxing the operation being applied at each step;

(b) short margin notes naming the reason for a step, not only the step itself;

(c) annotated examples kept and reused as a personal reference during independent practice.

IV. Showing All Work as a Habit of Mind

Showing all work asks a student to make an invisible process visible, writing down steps that could, in theory, stay in a student's head. A skipped step hides two very different situations from a teacher: a student who solved that step right without writing it, and a student who guessed. Full written work turns both situations into the same clear record, closing the gap between what a student did and what a teacher can see and respond to. A habit built around showing all work protects a student on a test, where partial credit depends on a visible trail alone connecting a problem to its answer.

(a) Writing every operation, including steps that feel obvious in the moment;

(b) full written work as the only reliable evidence of a correct process, separate from a correct final answer;

(c) partial credit as a direct, practical payoff of a visible work habit.

V. Writing Reflections on Mistakes

A reflection on a mistake asks a student to name what went wrong in specific terms, moving past a general feeling of frustration toward a precise diagnosis a student can act on. Naming and monitoring one's own thinking, the core of what researchers termed metacognition, gave a learner a way to catch and correct an error mid-process rather than after a grade came back (Flavell, 1979). A short written reflection, structured around what happened, why it happened, and what will change next time, turns an error from a source of shame into a specific plan. Reflections collected over a term build a personal error log that a student can review before a test, targeting the specific mistakes that history shows are most likely to recur.

(a) A three-part reflection structure: what happened, why it happened, what changes next time (Flavell, 1979);

(b) a personal error log built from reflections collected across a term;

(c) reflection treated as a specific, targeted plan rather than a general apology for a low score.

VI. The Math Journal as a Running Record

A math journal gives a student space to write about mathematics in ordinary language, connecting a new concept to a real example, a personal observation, or an honest question. Journal writing, including freewriting, learning logs, and short reflective entries, gave teachers direct insight into a student's reasoning and gave students a chance to process ideas at their own pace (Countryman, 1992). An entry does not need to be long to be valuable; three honest sentences about what confused a student today can guide an entire review session two weeks later. A journal maintained across a full year becomes a running record of growth, showing a student sentences from September that look nothing like the reasoning that same student shows in June.

(a) Short, regular entries connecting a new concept to a real example or personal observation;

(b) freewriting or learning-log formats that lower the pressure to write a perfect response (Countryman, 1992);

(c) a full-year record a student can reread to see their own reasoning develop over time.

Research Foundations Cited in This Chapter

Countryman, J. (1992). *Writing to learn mathematics: Strategies that work, K-12*. Heinemann.

Flavell, J. H. (1979). Metacognition and cognitive monitoring: A new area of cognitive-developmental inquiry. *American Psychologist*, *34*(10), 906–911.

Kiewra, K. A. (1989). A review of note taking: The encoding-storage paradigm and beyond. *Educational Psychology Review*, *1*(2), 147–172.

Pauk, W. (1962). *How to study in college*. Houghton Mifflin Company.

Supplemental Features for This Chapter

Math Mindset Moment: a short boxed feature asks students to write one honest sentence in their journal about a moment of confusion from today's lesson before leaving class. The feature closes with the line: "A confused sentence, written down, is the start of a clear one."

Student Success Story: a composite profile follows a student who treated notes as a place to copy answers until a Cornell-style summary section forced a written restatement in the student's own words. The story traces the specific unit test where that summary habit turned a familiar pattern of forgotten steps into a passing score built on visible, organized reasoning.

Reflection Questions:

(a) describe the difference between a page of notes you copied and a page of notes where writing helped you think;

(b) write a Cornell-style summary, in your own words, for the last math lesson you attended;

(c) choose one worked example from your notes and annotate it, marking the reasoning behind each step;

(d) write a three-part reflection on one recent mistake: what happened, why it happened, and what will change next time;

(e) write one journal entry connecting a current math concept to something from outside the classroom.

Chapter Takeaway

Writing functions as a tool for thinking in mathematics, not a chore that follows the real work of solving a problem. Cornell notes, annotation, full written work, error reflections, and a running journal each turn a different moment of a math lesson into visible evidence of a student's reasoning (Countryman, 1992; Flavell, 1979; Kiewra, 1989; Pauk, 1962). A student who writes across every lesson builds a record that supports memory during a test and supports confidence long after the test ends. That written record also becomes the raw material for a skill Chapter 6 addresses next: explaining a math process out loud, using the same reasoning a student has already practiced putting into words. Chapter 6 shifts from writing to speaking, teaching students how to talk about mathematics with the same clarity they have begun to develop on paper.

Chapter 6: Learning the Process, Not Just the Answer

Understanding the "Why" Behind Mathematics

Getting Started

Too often, students become focused on finding the correct answer without understanding how they arrived there. Mathematics is not about guessing correctly; it is about developing logical thinking and problem-solving skills. The process matters because life rarely presents situations with obvious answers. Learning how to analyze problems, identify patterns, choose appropriate strategies, and evaluate solutions prepares students for challenges far beyond the classroom.

Imagine a contractor building a house. The finished home is important, but every measurement, blueprint, foundation, and safety inspection must be completed correctly before construction is finished. Missing one important step could create serious problems later. Mathematics follows the same principle. If you're calculating how much paint is needed to cover the walls of a room, simply knowing the final number isn't enough. You must first determine the dimensions, calculate the area, subtract the windows and doors, and estimate the amount of paint required. Understanding the process ensures that your answer makes sense.

Whenever you solve a math problem, ask yourself, "Why did this method work?" and "Could I explain this to someone else?" These questions deepen understanding and help you become an independent thinker. The more you value the process over simply getting the answer, the more capable and confident you will become as a mathematician.

Real-Life Math Connection

You need to paint a bedroom.

Solution

Length = 12 ft

Width = 10 ft

Height = 8 ft

Area of four walls

2(12×8) + 2(10×8)

=

192 + 160

=

352 square feet

If one gallon covers 350 square feet, then 352 / 350 = 1.01

You'll need 2 gallons (or 1 gallon and 1 quart to be more cost-effective); otherwise, you'd be short of paint by 2 square feet, not counting wasted paint (on the brush, roller, roller tray, the sides of the can, the lid, the stir stick), which would mean you'd be short of paint by more than 2 square feet, maybe 3 or 4.

Understanding the process prevents costly mistakes.

Chapter Purpose and Guiding Question

Why does the ability to explain a mathematical process matter more than producing a correct answer alone?

Chapter 6 asks a harder question than "What is the answer," pushing students toward "why does this work" instead. A correct answer, reached through a guessed operation or a memorized shortcut, tells a teacher almost nothing about what a student understands, while a wrong answer paired with clear reasoning often reveals more mathematical thinking than the correct guess. Explanation-based study produced greater problem-solving skill than passive review, since students who explained worked examples to themselves built knowledge that transferred to new problems, not just the practiced one (Chi et al., 1989). A student who cannot explain a process has not yet finished learning it, regardless of how many correct answers that student produced on a worksheet. Chapter 6 treats reasoning, not the final number, as the actual object of instruction, a shift that changes what counts as a strong answer in a math classroom. Talking through mathematics, writing about mathematics, and working alongside another student who is also talking and writing all give reasoning a place to become visible instead of staying locked inside one student's head. The chapter builds five specific tools for that visibility: self-explanation, think-aloud strategies, structured discourse, peer collaboration, and written justification. Each tool asks the same underlying question in a different format: Can a student defend this answer, not just produce it? The remaining sections teach each tool and show how the tools reinforce one another across a single lesson.

Learning Objectives

Chapter 6 sets five learning outcomes for students to reach before its final page:

(a) explain, out loud, the reasoning behind one full solution rather than stating the answer alone;

(b) use a think-aloud strategy to narrate their thinking while solving an unfamiliar problem (Davey, 1983);

(c) participate in a structured math discussion using an accountable-talk sentence starter (Chapin et al., 2003);

(d) work through one problem with a partner, taking turns explaining and questioning each step (Webb, 1991);

(e) construct a written response that justifies a solution using evidence and mathematical vocabulary.

Section Outline

I. Answer-Getting vs. Sense-Making: Naming the Habit

Answer-getting treats a math problem as a single hurdle to clear, valuing speed and the correct final answer over the path taken to reach it. Sense-making treats the same problem as a question worth understanding, valuing a clear account of the reasoning as much as the number at the bottom of the page. A student trained toward answer-getting often cannot explain a correct solution the next day, since the answer was produced through pattern-matching rather than understanding. Naming this distinction for students gives them permission to slow down, since many students believe speed alone signals math skill.

(a) The visible signs of answer-getting: guessing, pattern-matching, and rushing past a check step;

(b) the visible signs of sense-making: pausing to explain, checking reasoning against the question asked;

(c) classroom norms that reward a clear explanation on the same level as a correct final answer.

II. The Self-Explanation Effect: Why Explaining Beats Reviewing

Self-explanation asks a student to stop and account for a step immediately after taking it, turning a worked example into an active object of study rather than a model to skim. Students who generated frequent, accurate self-explanations while studying worked examples built knowledge that reached beyond the specific problem in front of them, applying that knowledge to new, related problems with far greater success than students who studied the same examples without explaining them (Chi et al., 1989). A short prompt, such as "why did I subtract here" or "what rule allows this step," turns a moment of copying into a moment of real analysis. Teachers can build this habit into a lesson, pausing worked examples at key steps and asking a student to supply the missing explanation before moving forward.

(a) Self-explanation prompts inserted at key steps of a worked example;

(b) the gap between knowledge that transfers to new problems and knowledge tied to one memorized example (Chi et al., 1989);

(c) a habit of pausing after each step to ask "why," even when working alone.

III. Think-Aloud Strategies: Making Reasoning Audible

A think-aloud strategy asks a student, or a teacher modeling for students, to say every thought out loud while working through a problem, turning an invisible mental process into an audible one. Modeling a skilled thinker's process out loud, including false starts and self-corrections, gave learners access to strategies an expert uses without conscious comment, strategies a struggling student cannot copy if that student never hears them named (Davey, 1983). A teacher who thinks aloud while solving a problem shows students that hesitation, backtracking, and rechecking are part of strong mathematical thinking rather than signaling weakness. Students who practice their own think-alouds, even alone or with a partner, build the same monitoring habit that separates a strategic problem solver from one who guesses and hopes.

(a) Teacher modeling of a think-aloud, including hesitation, false starts, and self-correction (Davey, 1983);

(b) student think-alouds practiced alone, narrating each decision during a problem;

(c) partner think-alouds, where one student solves aloud while another listens for a specific reasoning move.

IV. Mathematical Discourse: Talk Moves That Build Understanding

Mathematical discourse turns a classroom from a room where one person talks and many listen into a room where reasoning gets built together through structured talk. Specific talk moves, including revoicing a student's idea, asking a student to restate a classmate's reasoning, and pressing for justification, gave teachers concrete tools for deepening a discussion beyond a simple right-or-wrong exchange (Chapin et al., 2003). A sentence starter such as "I agree with that because" or "I want to add onto what was said" gives a hesitant student a low-risk entry point into a conversation that might otherwise intimidate them. Discourse norms, taught in direct terms and practiced daily, turn class discussion into a rehearsal space for the same reasoning a student will later need to write down alone.

(a) Talk moves including revoicing, restating, and pressing for justification (Chapin et al., 2003);

(b) sentence starters that give every student, including a hesitant one, a low-risk way into a discussion;

(c) discourse norms taught and practiced across the term rather than assumed from the first day.

V. Peer Collaboration: Learning From and With Other Students

Peer collaboration puts two forms of reasoning to work at once, asking a student to solve a problem while also explaining the solution to someone who might see it differently. Giving and receiving help during small-group work related to mathematics learning in specific ways, with detailed, elaborated explanations supporting the learning of both the student giving help and the student receiving it, while brief or unexplained answers offered little benefit to either student (Webb, 1991). A student pushed to explain a step to a confused partner often notices a gap in their own reasoning that a silent, individual worksheet would never have exposed. Structured collaboration, with clear roles and a specific task, produces far more of this productive explaining than an unstructured group told only to work together.

(a) The difference in benefit between elaborated explanations and brief, unexplained answers during peer work (Webb, 1991);

(b) structured roles, such as explainer and questioner, that increase productive talk during collaboration;

(c) a partner's confusion used as a diagnostic tool for a student's own reasoning gaps.

VI. Constructing Written Responses: Proof on the Page

A constructed written response asks a student to defend a solution in full sentences, using evidence and vocabulary the way a mathematician would defend a claim in a paper. Writing a justification forces a level of precision that speaking alone can allow a student to skip, since a reader cannot fill in a missing step the way a listener sometimes fills in a pause. A strong written response names the strategy used, shows the key steps, and states the conclusion in a sentence that could stand on its own without the original problem attached. Constructed responses give students at every level, including a student who freezes during a class discussion, a private space to build and revise the same reasoning skill discourse and collaboration have already practiced out loud.

(a) A written response that names the strategy, shows key steps, and states a conclusion in full sentences;

(b) evidence and math vocabulary used the way a mathematician would use them to defend a claim;

(c) constructed responses as a private practice space for a student not yet comfortable speaking in front of a group.

Research Foundations Cited in This Chapter

Chapin, S. H., O'Connor, C., & Anderson, N. C. (2003). *Classroom discussions: Using math talk to help students learn*. Math Solutions Publications.

Chi, M. T. H., Bassok, M., Lewis, M. W., Reimann, P., & Glaser, R. (1989). Self-explanations: How students study and use examples in learning to solve problems. *Cognitive Science, 13*(2), 145–182.

Davey, B. (1983). Think aloud: Modeling the cognitive processes of reading comprehension. *Journal of Reading*, *27*(1), 44–47.

Webb, N. M. (1991). Task-related verbal interaction and mathematics learning in small groups. *Journal for Research in Mathematics Education*, *22*(5), 366–389.

Supplemental Features for This Chapter

Math Mindset Moment: a short boxed feature challenges students to solve one homework problem correctly and then write a deliberately flawed explanation of it, swapping papers with a partner to find the planted error. The feature closes with the line: "If you can explain it, you own it."

Student Success Story: a composite profile follows a student who could produce correct answers all year but froze whenever asked to explain a method out loud. The story traces the specific weeks of low-stakes think-aloud practice with a trusted partner that turned that silence into a clear, confident explanation during a final presentation.

Reflection Questions:

(a) explain one solution from a recent assignment out loud to another person, then write down what you said;

(b) describe a moment when explaining your thinking to a classmate helped you notice a mistake in your own reasoning;

(c) practice a think-aloud on one new problem, narrating every decision before you write a single step;

(d) use one sentence starter from this chapter to respond to a classmate's idea during your next math class;

(e) write a full justification for one answer, including the strategy used and the key steps that support it.

Chapter Takeaway

Reasoning, not a final number, functions as the true target of mathematics instruction, a shift this chapter asks both students and teachers to make together. Self-explanation, think-aloud strategies, structured discourse, peer collaboration, and written justification each give that reasoning a different outlet, and each outlet strengthens the others through repeated practice (Chapin et al., 2003; Chi et al., 1989; Davey, 1983; Webb, 1991). A student who can explain a process out loud, defend it in writing, and teach it to a confused partner has demonstrated an understanding no single correct answer could prove alone. That depth of understanding depends on a resource this chapter has assumed throughout: sustained attention held across an entire problem instead of scattered across a distracted room. Chapter 7 turns to focus itself, teaching students the concentration skills that make every strategy in this chapter possible to sustain.

Chapter 7: Focus Over Frustration

How Concentration Builds Mathematical Confidence

Getting Started

Every student experiences frustration while learning mathematics. The difference between successful students and struggling students is not intelligence: it is how they respond when challenges arise. Frustration often causes students to give up too quickly, while focused students learn to slow down, take a deep breath, and approach the problem one step at a time. Focus allows the brain to work through obstacles instead of becoming overwhelmed.

Think about a quarterback leading a football team during the final minutes of a close game. With thousands of fans cheering and time running out, losing focus for even a few seconds could change the outcome. Mathematics requires the same mental discipline. Consider balancing a monthly budget. You must carefully calculate income, subtract expenses, compare costs, and plan for savings. Losing focus can result in overdraft fees or unpaid bills. Careful attention to each calculation leads to better financial decisions and greater confidence.

When frustration appears, pause instead of quitting. Break large problems into smaller parts, eliminate distractions, and celebrate small victories along the way. Every difficult problem solved strengthens your perseverance. Focus is a skill that improves with practice, and every moment of concentration moves you one step closer to mastering mathematics.

Real-Life Math Connection

Solution

Monthly Income

$3,200

Bills

Rent $1,200

Utilities $300

Car Payment $450

Food $500

Savings $300

Money Left

$450

Remaining focused helps prevent overspending.

Chapter Purpose and Guiding Question

How can focus be trained as a skill that carries a student through a difficult mathematical problem?

Chapter 7 treats focus as a skill built through practice, not a trait some students bring to class, and others do not. A wandering mind during a math lesson costs more than a missed sentence; a wandering mind during a multi-step problem can undo three correct steps in the time it takes to glance at a phone. Attention operates through distinct brain networks responsible for alerting, orienting, and resolving conflict between competing pieces of information, networks that respond to training the same way a muscle responds to repeated use (Posner & Petersen, 1990). Math anxiety adds a second demand on top of an already limited attention budget, since an anxious mind spends part of its capacity monitoring the anxiety itself instead of the problem on the page (Ashcraft, 2002). A student asked to sit still and concentrate, without ever being taught how, faces an instruction with no method attached. Chapter 7 supplies that method: routines, brief mindfulness practices, and a clear account of what focus costs a distracted brain in real terms. Productive struggle, introduced in Chapter 2, depends on a mind able to stay with a hard problem long enough to work through it, thereby making focus the engine behind nearly every other strategy in this book (National Council of Teachers of Mathematics [NCTM],

2014). A student who learns to redirect a wandering mind gains a skill that outlasts any single math unit. The remaining sections build that skill through routine, practice, and repetition.

Learning Objectives

Chapter 7 sets five learning outcomes for students to reach before its final page:

(a) name the three attention networks involved in focused work: alerting, orienting, and conflict resolution (Posner & Petersen, 1990);

(b) explain how math anxiety competes with attention for the same limited mental capacity (Ashcraft, 2002);

(c) practice one focus-building routine at the start of a math lesson;

(d) use one brief mindfulness strategy to reset attention during a difficult problem (Mrazek et al., 2013);

(e) sustain focus across a full problem set, tracking where and why attention drifted.

Section Outline

I. Attention as a Trainable System, Not a Fixed Trait

Attention functions through a set of interacting brain networks rather than a single switch a student turns on or off. An alerting network keeps a student ready to receive information, an orienting network directs attention toward a specific detail, such as a negative sign, and a conflict-resolution network manages competing responses when two possible next steps both seem reasonable (Posner & Petersen, 1990). A student who struggles to focus is not missing a personality trait; that student is missing practice with one or more of these specific networks. Naming attention as a system with parts, rather than a single, vague quality, gives teachers a target for specific practice rather than a general command to concentrate.

(a) the alerting network, which keeps a student ready to receive new information;

(b) the orienting network, which directs attention toward a specific detail or step;

(c) the conflict-resolution network, which manages competing responses during a multi-step problem (Posner & Petersen, 1990).

II. When Anxiety Hijacks Attention: The Cost of a Distracted Mind

Anxiety and attention compete for the same limited resource, so a student worried about failing a quiz has less mental capacity left for the quiz itself. Working memory carried the double load of solving a problem and monitoring an anxious internal narrative simultaneously, a split that reduced accuracy and speed on math tasks in measurable ways (Ashcraft, 2002). A student who appears distracted during a test may not lack focus in a general sense; that student's focus may be full of worry instead of the problem in front of them. Recognizing this split gives a teacher a reason to address the anxiety head-on, since a student cannot out-practice a mind that has no spare attention left to practice with.

(a) The shared mental resource anxiety and attention both draw from (Ashcraft, 2002);

(b) the difference between a student who lacks focus and a student whose focus is filled with worry;

(c) classroom strategies that reduce anxiety first, before asking for more concentration.

III. Focus-Building Routines: Structuring the First Five Minutes

A consistent opening routine trains a student's attention system the same way a consistent warm-up trains a muscle before physical activity. A short, predictable task at the start of every math class, such as a single review problem or a brief mental math challenge, signals to the brain that focused work has begun and pulls attention away from whatever occupied it moments before. Routines succeed through repetition, not novelty, since a brain expends less energy on a familiar sequence and can direct more of that saved energy toward the math problem itself. A missing routine forces a student to spend the first several minutes of every class settling into the room, time that a consistent structure would have reclaimed for instruction.

(a) A short, predictable opening task repeated at the start of every class;

(b) routines built on repetition rather than novelty, to reduce the mental cost of getting started;

(c) a clear signal, verbal or visual, marking the shift from transition time to focused work time.

IV. Mindfulness Strategies for the Math Classroom

Mindfulness strategies train a student to notice a wandering mind and return attention to the present task, a skill with clear relevance to a subject in which one missed line can derail an entire solution. Short mindfulness-based practices integrated into a school day produced a measurable positive effect on student well-being and behavior across dozens of studies and thousands of students (Zenner et al., 2014). A brief training period improved both working memory capacity and performance on a demanding cognitive test, while reducing the frequency of mind wandering reported during the task (Mrazek et al., 2013). A two-minute breathing exercise before a test or a brief body scan before an independent work block gives a student a concrete tool to redirect attention rather than a vague instruction to calm down.

(a) Short breathing exercises used before a quiz, test, or independent work block;

(b) school-based mindfulness programs and their measured effect on student well-being and behavior (Zenner et al., 2014);

(c) brief mindfulness training linked to improved working memory and reduced mind wandering during demanding tasks (Mrazek et al., 2013).

V. Building Endurance: Sustaining Focus Across a Full Problem Set

Focus for thirty seconds and focus for thirty minutes draw on related but distinct skills, and a classroom that only ever asks for the first never builds the second. Endurance grows the way any trained capacity grows: through practice sessions that extend in length over time, paired with a clear signal for when a short reset is allowed. A student who tracks the exact point where attention drifted during a problem set, rather than just noting that they got distracted, builds the kind of specific self-awareness that makes the next attempt easier to sustain. Short, permitted breaks between problem sets better protect endurance than a rule demanding constant, unbroken focus, since a brief reset prevents the kind of fatigue that leads to a much longer lapse later. Even marathon runners take water breaks, and nobody accuses them of quitting.

(a) Practice sessions that extend in length over time rather than jumping straight to a full period;

(b) a clear signal for permitted short resets during longer work blocks;

(c) specific self-tracking of where and why attention drifted, rather than a general note of distraction.

VI. Productive Struggle Revisited: Focus as the Engine Behind It

Productive struggle, introduced in Chapter 2 as a condition for deep learning, requires a mind able to stay with confusion instead of escaping it at the first sign of difficulty. Teacher actions that support productive struggle, including giving students time and resisting the urge to remove a challenge too soon, depend on students having enough sustained focus to use that time well (NCTM, 2014). A student with strong focus skills can sit inside a hard problem long enough for productive struggle to do its work; a student without those skills abandons the problem before struggle has a chance to become productive. Chapter 7 closes with a link between focus and every strategy already covered in this book, from self-explanation to error reflection, since none of those strategies work without a mind able to stay in the room.

(a) The dependence of productive struggle on sustained attention (NCTM, 2014);

(b) the difference between abandoning a problem early and working through it with focus intact;

(c) focus as a prerequisite skill underneath self-explanation, discourse, and written justification from Chapter 6.

Research Foundations Cited in This Chapter

Ashcraft, M. H. (2002). Math anxiety: Personal, educational, and cognitive consequences. *Current Directions in Psychological Science*, *11*(5), 181–185.

Mrazek, M. D., Franklin, M. S., Phillips, D. T., Baird, B., & Schooler, J. W. (2013). Mindfulness training improves working memory capacity and GRE performance while reducing mind wandering. *Psychological Science*, *24*(5), 776–781.

National Council of Teachers of Mathematics. (2014). *Principles to actions: Ensuring mathematical success for all.* NCTM.

Posner, M. I., & Petersen, S. E. (1990). The attention system of the human brain. *Annual Review of Neuroscience*, *13*, 25–42.

Zenner, C., Herrnleben-Kurz, S., & Walach, H. (2014). Mindfulness-based interventions in schools: A systematic review and meta-analysis. *Frontiers in Psychology*, *5*, Article 603.

Supplemental Features for This Chapter

Math Mindset Moment: a short boxed feature walks students through a sixty-second breathing routine to use before a quiz begins, paired with the line: "My focus is a skill, and I am training it right now."

Student Success Story: a composite profile follows a student who lost significant time on every test re-reading the same line without absorbing it, a pattern traced back to test-day anxiety rather than a reading skill gap. The story traces the specific weeks of breathing practice and a fixed opening routine that closed the gap between what the student knew and what the student could demonstrate on a test.

Reflection Questions:

(a) describe one moment this week when your mind wandered during math class and name what pulled your attention away;

(b) practice a sixty-second breathing exercise before your next math assignment and describe how your focus felt afterward;

(c) explain the difference between a mind that lacks focus and a mind that is full of worry instead of the problem at hand;

(d) design one opening routine you could use at the start of every math class to signal that focused work has begun;

(e) track where your attention drifted during one full problem set and name one pattern you noticed.

Chapter Takeaway

Focus functions as a trainable system built from specific networks, rather than as a single trait a student either has or lacks (Posner & Petersen, 1990). Math anxiety competes with that system for the same limited attention, which means reducing anxiety and building focus work as two parts of the same project rather than two separate ones (Ashcraft, 2002). Routines, brief mindfulness practices, and deliberate endurance-building each strengthen a different part of that system, giving a teacher more than one entry point for the same goal (Mrazek et al., 2013; Zenner et al., 2014). Productive struggle, the condition this book has returned to since Chapter 2, depends on every one of these focus skills working together in the room at once (NCTM, 2014). Chapter 8 turns to what happens after focus fails and a mistake occurs, teaching students to treat that mistake as data rather than a verdict on their ability.

Chapter 8: Mistakes Are Part of Mastery

Turning Errors into Learning Opportunities

Getting Started

Many students fear making mistakes because they believe mistakes are signs of failure. In reality, mistakes are evidence that learning is taking place. Every mathematician, scientist, engineer, and inventor has made countless errors before discovering successful solutions. Mistakes reveal misconceptions, highlight areas for growth, and provide opportunities to deepen understanding. Instead of avoiding mistakes, successful learners examine them carefully and use them as stepping stones toward improvement.

Imagine a basketball player practicing free throws. Missing several shots during practice does not mean the player is incapable of becoming an excellent shooter. Each missed shot provides valuable feedback about balance, form, and technique. Mathematics works the same way. If you incorrectly calculate the amount of interest earned on a savings account, reviewing your work helps you understand where your thinking went wrong. That correction prepares you to make better financial decisions in the future.

Adopt the mindset that every mistake teaches an important lesson. Ask yourself what happened, why it happened, and how you can avoid making the same error again. Progress is not measured through perfection but through continuous improvement. Every corrected mistake builds stronger understanding, greater confidence, and a more resilient mathematical mindset.

Real-Life Math Connection

Imagine you calculated a 20% tip on a $50 meal.

Incorrect answer:

$20

Review:

20%

=

0.20

$0.20 \times 50 = \$10$

Finding the mistake teaches more than simply getting the answer right.

Chapter Purpose and Guiding Question

How can a wrong answer serve as a source of information rather than a verdict on a student's ability?

Chapter 8 asks students and teachers to treat a wrong answer as information rather than as a verdict on ability, a distinction the whole book has built toward since Chapter 1. A mistake, examined with care, often contains a more precise diagnosis of a gap in understanding than a correct answer could ever supply, since a correct answer can hide a shaky process behind a lucky result. A growth mindset changed how the brain responded to an error at the neural level, producing heightened attention to the mistake and greater accuracy on the very next attempt (Moser et al., 2011). Many classrooms still treat an error as an endpoint, marked wrong and left behind, when that same error could function as the starting point for a specific, targeted inquiry (Borasi, 1994). Feedback that named a clear next step, rather than only a grade or a vague comment, produced far stronger learning gains, which means how a teacher responds to a mistake matters as much as the mistake itself (Hattie & Timperley, 2007). Chapter 8 gives teachers and students a shared vocabulary for handling errors: analysis instead of embarrassment, feedback instead of judgment, reflection instead of avoidance. A classroom that normalizes this vocabulary changes what a student risks when attempting a hard problem in front of others. The remaining sections build that vocabulary through specific classroom practices, connecting back to the growth mindset foundation laid in Chapter 2.

Learning Objectives

Chapter 8 sets five learning outcomes for students to reach before its final page:

(a) explain the difference between treating a mistake as a verdict and treating it as data (Moser et al., 2011);

(b) analyze one incorrect solution to identify the specific step where the error occurred and why;

(c) give a classmate feedback that names a clear next step, not just a right or wrong judgment (Hattie & Timperley, 2007);

(d) write a brief reflection connecting one recent mistake to a specific change in strategy (Flavell, 1979);

(e) describe one classroom norm that makes it safer to share a wrong answer out loud.

Section Outline

I. Reframing the Mistake: From Verdict to Data

A verdict closes a question; data opens one, and the difference between those two responses to a wrong answer shapes whether a student tries again or shuts down. Treating a mistake as a verdict tells a student something fixed about their ability, while treating the same mistake as data tells a student something specific about their next step. A student handed a paper marked only with an X has received a verdict; a student handed a paper marked with a question pointing at the exact site of the error has received data. Chapter 8 asks teachers and students to practice noticing which response a classroom, a grade, or an inner voice is offering after every mistake.

(a) The verdict response: “I got it wrong, so I am bad at math”;

(b) the data response: “I got it wrong here, so I need to check this specific step”;

(c) classroom and self-talk habits that push a student toward data instead of a verdict.

II. The Neuroscience of Errors: What Happens in the Brain After a Mistake

The brain produces a measurable electrical response within a fraction of a second after a person makes an error, a signal tied to the amount of attention the error receives. Students holding a growth mind-set showed a stronger version of this error-related brain response and greater accuracy on the very next trial after a mistake, compared with students holding a fixed mind-set (Moser et al., 2011). A brain that pays close attention to an error, rather than looking away from it, gathers more of the information that error has to offer. Teaching students this specific mechanism, rather than a vague call to "learn from your mistakes," gives the advice a concrete, physical basis a skeptical student can respect.

(a) The brain's rapid electrical response to an error and its link to attention (Moser et al., 2011);

(b) the connection between a growth mind-set and stronger post-error accuracy;

(c) explaining this brain research to students as concrete evidence rather than just encouragement.

III. Error Analysis as a Classroom Practice

Error analysis asks a student to become a detective, investigating their own work and searching for the exact point where a correct process turned into an incorrect one. Errors, used as a deliberate starting point for inquiry rather than only as a problem to fix, opened space for students to explore the mathematical ideas underlying a mistake in greater depth than a straightforward correct answer would have prompted (Borasi, 1994). A class period built around analyzing a single rich error, worked through together on the board, teaches more about a common misconception than ten practice problems solved the right way ever could. Structured error analysis gives every student practice at a skill working mathematicians use daily: checking a claim, finding the flaw, and repairing the reasoning.

(a) A single rich error analyzed as a whole-class activity rather than corrected in isolation;

(b) errors used as a starting point for inquiry into the mathematical idea behind them (Borasi, 1994);

(c) a repeatable process: locate the flaw, explain the flaw, repair the reasoning.

IV. Productive Feedback: What Helps and What Doesn't

Not all feedback helps a student learn, and some feedback works against learning even when it feels supportive in the moment. Feedback aimed at the task and the process, answering where a student is going, how that student is doing, and what comes next, produced far stronger gains than feedback aimed only at the person, such as general praise for being smart (Hattie & Timperley, 2007). A grade alone, without a specific comment attached, functions closer to a verdict than to feedback, since a number carries no information about what to do another way next time. Teachers who build feedback around these three questions give students a next step to act on instead of a score to accept or reject.

(a) The three feedback questions: where am I going, how am I doing, what comes next (Hattie & Timperley, 2007);

(b) task- and process-focused feedback compared with praise aimed only at the person;

(c) a grade alone as an incomplete form of feedback, missing a specific next step.

V. Growth Mindset Practices Applied to Error Culture

A growth mindset gives a classroom a shared belief system; error culture gives that belief system a daily practice, turning an abstract idea into a lived classroom habit. A teacher who works through their own mistake in front of the class, thinking aloud through the correction, models the exact behavior error culture asks of students. Language choices reinforce one mindset or another through repeated exposure, meaning the specific words a teacher uses in response to a wrong answer either build or erode a classroom's error culture over a semester (Dweck, 2006). A classroom that celebrates a well-analyzed mistake with the same visibility it gives a correct answer teaches students that both outcomes belong to the same learning process.

(a) Teacher modeling of their own mistakes, including the correction process, done in front of the class;

(b) specific language choices that reinforce a growth-oriented error culture over time (Dweck, 2006);

(c) visible celebration of well-analyzed mistakes alongside correct answers.

VI. Building a Classroom Where Mistakes Are Normalized

A normalized error culture does not happen through a single poster on a wall; it happens through repeated, consistent classroom routines that make sharing a wrong answer feel ordinary rather than risky. Reflective writing after a mistake, connecting what happened, why it happened, and what changes next, builds metacognitive awareness a student can carry into the next problem (Flavell, 1979). A "mistake of the day" routine, where one anonymous or volunteered error becomes the subject of whole-class analysis, gives error culture a predictable home inside every lesson instead of leaving it to chance. A classroom that has normalized mistakes in a real way produces students who raise their hands with a wrong answer about as often as they raise a hand with a right one.

(a) A recurring "mistake of the day" or similar routine built into regular instruction;

(b) reflective writing that connects a mistake to a specific change in strategy (Flavell, 1979);

(c) classroom culture judged on whether students volunteer wrong answers with the same comfort as right ones.

Research Foundations Cited in This Chapter

Borasi, R. (1994). Capitalizing on errors as "springboards for inquiry": A teaching experiment. *Journal for Research in Mathematics Education*, *25*(2), 166–208.

Dweck, C. S. (2006). *Mindset: The new psychology of success.* Random House.

Flavell, J. H. (1979). Metacognition and cognitive monitoring: A new area of cognitive-developmental inquiry. *American Psychologist*, *34*(10), 906–911.

Hattie, J., & Timperley, H. (2007). The power of feedback. *Review of Educational Research*, *77*(1), 81–112.

Moser, J. S., Schroder, H. S., Heeter, C., Moran, T. P., & Lee, Y. H. (2011). Mind your errors: Evidence for a neural mechanism linking growth mind-set to adaptive posterror adjustments. *Psychological Science*, *22*(11), 1484–1489.

Supplemental Features for This Chapter

Math Mindset Moment: a short boxed feature asks students to find and correct one "planted" error in a sample solution, then explain in one sentence how they found it. The feature closes with the line: "A mistake caught is proof that I am paying attention."

Student Success Story: a composite profile follows a student who once erased every wrong answer before anyone could see it, until a daily "mistake of the day" routine made sharing an error a normal, even valued, part of class. The story traces the specific class period when that student first volunteered an incorrect answer, an error that led the whole class to a breakthrough in understanding a common misconception.

Reflection Questions:

(a) describe one recent mistake and explain the exact step where your process went wrong;

(b) explain the difference between feedback that told you an answer was wrong and feedback that told you what to try next;

(c) practice explaining a mistake out loud to a partner, treating it as a clue rather than a failure;

(d) describe one classroom habit that would make you more willing to share a wrong answer out loud;

(e) rewrite one piece of feedback you have received into a version that names a specific next step.

Chapter Takeaway

Mistakes function as data that a student can use, not as a verdict that a student must accept, and the brain itself supports this reframing at a physical level (Moser et al., 2011). Error analysis, productive feedback, and growth mindset language each turn an individual mistake into a resource for the whole room (Borasi, 1994; Dweck, 2006; Hattie & Timperley, 2007). Reflective writing closes the loop, turning a single moment of confusion into a specific, actionable plan a student can carry forward (Flavell, 1979). A classroom built around this vocabulary changes what a wrong answer costs a student in social terms, not only in academic ones. Chapter 9 turns to that social cost, addressing the peer pressure that leads some students to hide a correct answer rather than risk a wrong one in front of their classmates.

Chapter 9: Peer Pressure and Performance

Why Some Students Hide Their Intelligence

Getting Started

As students grow older, they often become more concerned about fitting in than standing out. Unfortunately, peer pressure can have a powerful influence on academic performance, especially in mathematics. Some students pretend not to care about their grades because they fear being labeled as "too smart" or "trying too hard." Others remain silent during class discussions because they are afraid of giving the wrong answer in front of their classmates. However, true confidence comes from pursuing excellence, not popularity. Every successful person eventually learns that others' opinions should never determine the value of education.

Consider a student who enjoys basketball and dreams of playing professionally. That student willingly spends hours practicing free throws, dribbling drills, and conditioning, even when friends choose to spend their time differently. The same commitment is required in mathematics. Imagine two friends preparing for a math test on percentages. One decides to study while the other chooses to play video games all evening. The student who studies learns how to calculate restaurant tips, determine discounts during shopping, and compare loan interest rates. Skills like these extend far beyond the classroom and become valuable throughout life. Temporary popularity may feel rewarding today, but knowledge creates opportunities that last a lifetime.

Surround yourself with people who encourage your growth rather than discourage your effort. Find classmates who are willing to study together, ask thoughtful questions, and celebrate academic success. Remember that your greatest competition is not the student sitting beside you: it is the person you were yesterday. Every decision to learn, improve, and persevere moves you closer to becoming the best version of yourself.

Real-Life Math Connection

Restaurant Bill

$82

Solution

Tip

18%

Tip

= 0.18 × 82

= $14.76

Total Bill

$96.76

Learning percentages prepares students for adulthood far beyond classroom tests.

Chapter Purpose and Guiding Question

How does peer pressure shape a student's willingness to show mathematical effort, and what can counter that pressure?

Chapter 9 names a pressure many students have felt but few have heard described out loud: the fear that visible academic effort will cost them standing with their peers. Some students learn early that raising a hand with a correct answer carries a social price, a burden that can shape real classroom behavior, from silence during discussion to a grade lowered on purpose. That price is neither universal nor inevitable, but where it exists, it shapes a student's choices in ways an outside observer rarely sees. Individual agency, school influence, and community forces combine to shape whether a student experiences math class as a space for growth or a site of social risk. Chapter 9 treats this pressure as a solvable problem rather than a fixed feature of a student's environment, drawing on evidence that identity, representation, and community can each work in the opposite direction, toward confidence instead of concealment. A brief intervention addressing students' sense of belonging on campus narrowed a three-year achievement gap, providing evidence that belonging is something a school can build, not just something a student either has or lacks (Walton & Cohen, 2011). Students who see people they can identify with succeeding in advanced math and STEM fields build a different mental model of who belongs in those rooms, a model this chapter treats as a resource worth building with intention. The chapter closes with a direct challenge: academic excellence deserves to be pursued without apology, and a peer group worth keeping will make room for that pursuit.

Learning Objectives

Chapter 9 sets five learning outcomes for students to reach before its final page:

(a) explain how the fear of social judgment can influence a student's classroom participation;

(b) describe how school influence, community forces, and individual agency combine to shape a student's math identity;

(c) identify one mathematician, scientist, or engineer and describe their path into the field;

(d) explain how a sense of belonging functions as a specific, buildable classroom practice rather than a fixed feeling (Walton & Cohen, 2011);

(e) describe one action a student could take to support a classmate's academic confidence rather than undermine it.

Section Outline

I. The Hidden Cost of Visible Intelligence

A correct answer, spoken out loud in front of classmates, can feel like a straightforward win in one classroom and a social risk in another, depending on the unwritten rules of that room. Some students learn to hide a correct answer, slow down on purpose, or stay silent rather than risk a label that separates them from their friends. Some students read academic effort as a betrayal of group identity, a burden tied to the social dynamics of their peer group rather than to any personal weakness. Naming this cost in plain terms, instead of pretending it does not exist, gives students and teachers a shared starting point for addressing it openly.

(a) The specific behaviors that signal hidden intelligence: slowing down, staying silent, downplaying a correct answer;

(b) the social dynamics behind the pressure some students feel to distance themselves from visible academic effort;

(c) the difference between a student who lacks confidence and one who is incurring a real social cost.

II. Identity and Achievement: When Group Belonging and Academic Effort Collide

Group identity and academic achievement do not have to compete, but in some environments, a student experiences them as though they must choose between them. School influence, community forces, and individual agency work together to shape whether a student experiences advanced math as an extension of their identity or a threat to it. A student whose peers, family members, and teachers treat academic excellence as a natural part of their identity seldom faces this false choice in the first place. Chapter 9 argues that the choice itself is the problem to solve, not the student who feels forced to make it.

(a) The false choice between group belonging and academic achievement;

(b) school influence, community forces, and individual agency as three forces shaping math identity;

(c) environments where academic excellence and identity reinforce each other rather than compete.

III. Belonging as an Intervention, Not Just a Feeling

Belonging sounds like an emotion a student either feels or does not feel, but researchers have shown it can serve as a specific, teachable target for intervention. A brief exercise helping college students understand that early struggles with belonging are common and temporary, rather than a sign that they do not belong, produced measurable gains in grades and well-being over the following three years (Walton & Cohen, 2011). The intervention worked because it addressed a specific belief, not a vague mood, giving students a concrete reason to reinterpret a hard week rather than read it as proof they did not belong. Classrooms can build smaller versions of this same message into everyday practice, normalizing struggle as a shared experience rather than a private, isolating one.

(a) Belonging treated as a specific belief that can shift, not a fixed personality trait;

(b) the measurable academic and health effects of a brief belonging intervention (Walton & Cohen, 2011);

(c) classroom practices that normalize early struggle as common and temporary rather than a personal verdict.

IV. Representation in STEM: Seeing Is Believing

A student's mental picture of who becomes a mathematician, scientist, or engineer shapes what that student believes is possible for themselves, often without either person ever saying a word about it. Structured programs that surround students with a community of high-achieving peers and mentors in the sciences produce high rates of advanced degree completion, well above those reported for students in comparable settings without such a structured community. Representation works through more than inspiration alone; it supplies a detailed, specific model of the exact steps a path into a field requires. A classroom that features mathematicians, scientists, and engineers from a range of backgrounds as a normal part of instruction, not a special unit reserved for one month a year, builds this model all year long instead of during one set-aside unit.

(a) Structured peer and mentor communities linked to strong STEM outcomes;

(b) representation as a source of specific, detailed models, not only general inspiration;

(c) mathematicians, scientists, and engineers from a range of backgrounds featured as a routine part of instruction throughout the year.

V. Cultural Confidence: Excellence Without Apology

Cultural confidence asks a student to hold their identity and academic excellence in the same hand, rejecting the idea that one requires the other to shrink. Culturally relevant pedagogy asks teachers to affirm a student's cultural identity while holding high academic expectations, treating the two goals as reinforcing rather than competing (Ladson-Billings, 1995). A student who has internalized cultural confidence can walk into an advanced math class and experience that space as an extension of who they already are, not a room where they must set part of themselves aside at the door. Teachers who model this stance, naming students' full identities with respect while expecting rigorous work, give students daily proof that the two goals were never in conflict to begin with.

(a) Cultural identity and academic excellence held together, not traded off against each other;

(b) culturally relevant pedagogy as a framework combining identity affirmation with high expectations (Ladson-Billings, 1995);

(c) teacher language that affirms identity while naming rigorous expectations in the same breath.

VI. Positive Peer Influence and Academic Leadership

A peer group can enforce silence with the same ease as it enforces excellence, and the direction it takes depends on which students step into leadership within it. A single confident student, willing to answer a hard question or defend a correct but unpopular answer, can shift the norms of an entire classroom faster than any adult intervention working alone. Structured peer communities built around a shared academic identity, rather than left to form on their own, provide students with both academic support and social protection. Chapter 9 asks students to consider academic leadership as a real option: someone willing to be first, so that the cost of visible intelligence drops for every student who follows.

(a) The peer group's power to enforce either silence or excellence, depending on its norms;

(b) one confident student as a norm-shifting force within a classroom;

(c) structured peer communities that provide both academic support and social protection.

Research Foundations Cited in This Chapter

Ladson-Billings, G. (1995). Toward a theory of culturally relevant pedagogy. *American Educational Research Journal*, *32*(3), 465–491.

Walton, G. M., & Cohen, G. L. (2011). A brief social-belonging intervention improves academic and health outcomes of minority students. *Science*, *331*(6023), 1447–1451.

Supplemental Features for This Chapter

Math Mindset Moment: a short boxed feature asks students to name one person, real or historical, who represents the kind of mathematician or scientist they could become, then write one sentence about what that person's path required. The feature closes with the line: "My excellence does not need anyone's permission."

Student Success Story: a composite profile follows a student who spent two years giving wrong answers on purpose to avoid standing out, until a peer-led study group made visible effort the norm rather than the exception inside that specific friend group. The story traces the semester when that student began answering first in class, a shift that also changed how classmates around them behaved.

Reflection Questions:

(a) describe a moment when you held back a correct answer or a strong effort because of how you thought others would react;

(b) name one person, from your community or from history, whose academic or professional path you admire and explain why;

(c) describe what a classroom would look and feel like if visible intelligence carried no social cost at all;

(d) explain one way you could support a classmate's confidence the next time that classmate answers a question in class;

(e) write one sentence you could say to yourself the next time you feel pressure to hide your ability.

Chapter Takeaway

The pressure to hide intelligence traces back to real social dynamics, not to any weakness in the students who feel it. Belonging, representation, cultural confidence, and peer leadership each provide a school with a specific, buildable tool for shifting that pressure in the other direction, toward a norm in which visible excellence carries pride rather than risk (Ladson-Billings, 1995; Walton & Cohen, 2011). None of these tools depend on a single heroic student changing an entire school alone; each one works through structure, repetition, and community built over time. A student who commits to academic leadership changes the room they are in, and a room changed enough times becomes a school with a different norm. Chapter 10 turns to the educators who intentionally build that norm, examining the specific classroom practices that make a math room safe enough for every student to bring their full effort and their full identity at once.

Chapter 10: Teachers Who Transform Math Culture

Creating Safe and Empowering Math Classrooms

Getting Started

Behind every confident math student is often a teacher who believed in them before they believed in themselves. Great mathematics teachers do much more than explain formulas and assign homework, inspiring curiosity, encouraging perseverance, and creating classrooms where mistakes are viewed as opportunities for growth rather than reasons for embarrassment. A transformational teacher understands that every student learns differently and recognizes that patience, encouragement, and high expectations can change the course of a child's life.

The best teachers also help students connect mathematics to the real world. Instead of teaching equations in isolation, they show students how architects use algebra to design bridges, how construction workers use geometry to measure buildings, and how statisticians use statistics to study diseases and improve patient care. Imagine a teacher asking students to design a small community park using geometric measurements. Students calculate the area needed for walking paths, playground equipment, basketball courts, and green spaces while staying within a construction budget. Suddenly, mathematics becomes meaningful because students see how numbers solve real problems.

Students should never hesitate to ask questions or seek additional help. Teachers appreciate effort and curiosity far more than perfection. Likewise, teachers must continue creating classrooms where every student feels respected, challenged, and capable of success. A single encouraging conversation from a caring teacher can change a student's confidence forever. Sometimes the greatest lesson a teacher gives is not how to solve an equation but how to believe in oneself.

Real-Life Math Connection

Designing a Community Park

Solution

Total land

100 ft × 80 ft

Area

8,000 square feet

Basketball court

4,700 sq ft

Playground

1,200 sq ft

Walking trail

900 sq ft

Remaining Green Space

1,200 sq ft

Students discover how geometry influences city planning.

Chapter Purpose and Guiding Question

What specific practices allow a teacher to build a classroom where every student can bring full effort without fear?

Chapter 10 turns from student practice to teacher practice, naming the specific moves that build a math classroom where every student can bring their full effort without fear. A student's confidence responds to the room they are taught in, not only to the strategies covered in earlier chapters, which makes the teacher one of the most powerful variables in this entire book. High expectations alone can tip into pressure without support attached, and support alone can tip into low rigor without expectations attached; the two together, held at the same time, produced the strongest outcomes for students (Ladson-Billings, 1995). Teacher expectations shaped student outcomes through a documented mechanism, not through vague hope: children whose teachers expected more from them, for reasons unrelated to actual ability, showed measurable gains in performance over the following year (Rosenthal & Jacobson, 1968). The quality of an early teacher-child relationship predicted a student's academic and behavioral trajectory for years afterward, indicating that relationship-building functions as instruction rather than a separate, optional add-on (Hamre & Pianta, 2001). Chapter 10 gives teachers concrete practices across five areas: expectations paired with support, culturally responsive teaching, trust-building, instructional clarity, and differentiated instruction. None of these practices require abandoning rigor; each one channels rigor through a relationship strong enough to carry it. The remaining sections give teachers specific classroom moves, not abstract philosophy, for building that kind of room.

Learning Objectives

Chapter 10 sets five learning outcomes for educators to reach before its final page:

(a) explain why high expectations without support, or support without high expectations, both undermine student achievement (Ladson-Billings, 1995);

(b) describe the mechanism through which teacher expectations influenced measured student performance (Rosenthal & Jacobson, 1968);

(c) identify two relationship-building practices linked to long-term student outcomes (Hamre & Pianta, 2001);

(d) design one differentiated task offering multiple entry points for the same math concept (Tomlinson, 2001);

(e) describe one culturally responsive teaching practice appropriate for a specific math unit (Gay, 2000).

Section Outline

I. High Expectations With Support: The Combination That Works

High expectations tell a student a teacher believes they can succeed; support tells a student how to get there, and a classroom needs both messages delivered together, not one instead of the other. A teacher who demands rigor without scaffolding produces anxiety rather than achievement, while a teacher who offers warmth without rigor produces comfort rather than growth. Effective teachers of students from cultural backgrounds outside the mainstream combined warm, personal relationships with unapologetic, high academic demands, refusing to treat the two as a trade-off (Ladson-Billings, 1995). A single sentence can carry both messages at once: "This is hard, and I am not leaving until you can do it."

(a) High expectations paired with visible scaffolding, not high expectations alone;

(b) warmth and rigor held together rather than treated as opposite ends of a scale;

(c) specific teacher language that communicates both belief and support in the same sentence (Ladson-Billings, 1995).

II. Culturally Responsive Teaching in the Math Classroom

Culturally responsive teaching treats a student's cultural background as a resource for instruction, not as a barrier to be worked around or a topic reserved for a single unit. Teaching filtered through a student's own cultural experiences and frames of reference produced stronger achievement across multiple measures for students whose cultural background differed from the classroom's dominant culture, compared with instruction that ignored those frames altogether (Gay, 2000). A math classroom applies this principle through concrete choices: word problems drawn from familiar contexts, examples that reflect a range of names and communities, and a teacher who treats a student's home language and experience as an asset in the room. Culturally responsive teaching and high academic expectations reinforce each other, since a student who feels recognized in a classroom brings more of their full effort to the rigorous work happening inside it.

(a) Instruction filtered through students' cultural experiences and frames of reference (Gay, 2000);

(b) word problems and examples that reflect familiar names, contexts, and communities;

(c) a student's home language and experience treated as an asset rather than an obstacle.

CRP rests on three commitments: academic success for every student, cultural competence that treats a student's background as a strength rather than a barrier, and an awareness of how the wider world shapes a student's access to opportunity (Ladson-Billings, 1994). A teacher who holds all three does more than swap names in a word problem. That teacher builds a room where a student's identity and math ability grow together rather than competing for space.

CRP does not implement itself just because a teacher believes in it. Programs that adopt it unevenly, without real training or follow-up, tend to produce uneven results: some classrooms transform, others barely change. A teacher new to this approach benefits from starting with one unit and one set of examples rather than trying to rebuild an entire course at once.

Some schools pair CRP with adaptive software such as ALEKS, which lets a teacher swap a generic word problem for one built around a student's own community: a recipe problem centered on a family's Sunday dinner, a geometry problem centered on the sloped roof of a traditional Navajo hogan. The tool only delivers on that promise when a teacher uses it that way. Software does not replace a teacher's judgment about which example will land with a specific group of students.

III. Building Trust: The Relationship Behind the Instruction

Trust functions as the foundation underneath every other strategy in this book, since a student will not risk a wrong answer, a question, or a moment of confusion in front of a teacher they do not trust. The quality of an early relationship between a teacher and a student predicted that student's academic and behavioral outcomes for years into the future, an effect that held even after accounting for a student's earlier skills and behavior (Hamre & Pianta, 2001). Trust builds through small, repeated actions: remembering a detail from a student's life, following through on a promise, and responding to a mistake with patience instead of frustration. A teacher who invests in relationship-building is not spending time that belongs to instruction; that teacher is building the exact condition instruction depends on to work.

(a) Small, repeated actions that build trust over time, rather than one large gesture;

(b) the long-term academic effect of an early, positive teacher-student relationship (Hamre & Pianta, 2001);

(c) a calm, patient response to a student's mistake as a trust-building act in itself.

IV. Student Engagement: Instructional Clarity as a Form of Respect

Engagement grows from clarity, not entertainment; a student stays with a lesson they can follow and drifts away from one that leaves them guessing at the goal. An objective stated in clear terms, a model worked through one step at a time, and a chance to practice with support each remove a specific barrier to engagement before that barrier can become disengagement. Clear instruction functions as a form of respect, since a teacher who explains a concept with care is telling a student their understanding matters enough to work for. Engagement built on clarity reaches more students than engagement built on novelty alone, since novelty fades while a clear structure keeps working lesson after lesson.

(a) An objective stated in clear terms, posted and referenced throughout a lesson;

(b) a model worked through one step at a time before independent practice begins;

(c) clarity treated as a form of respect for a student's time and effort.

V. Differentiated Instruction: One Classroom, Many Entry Points

A single lesson delivered one way to an entire class assumes a uniform readiness level that seldom exists in a real room. Differentiated instruction adjusts content, process, or product in response to a student's readiness, interest, or learning profile, giving every student an appropriate entry point into the same core concept (Tomlinson, 2001). A tiered task, offering the same underlying math concept at two or three levels of difficulty, lets a teacher hold every student to grade-level thinking without requiring them to start from the same place. Differentiation does not lower expectations; it removes an unnecessary obstacle standing between a student's current skill and the high expectation the classroom holds for that student.

(a) Adjusting content, process, or product in response to readiness, interest, or learning profile (Tomlinson, 2001);

(b) tiered tasks offering the same core concept at multiple levels of difficulty;

(c) differentiation understood as a path to high expectations, not a lowering of them.

VI. Becoming a Culture-Transforming Teacher

A teacher does not transform a classroom culture through one dramatic lesson; that teacher transforms it through hundreds of small, consistent choices repeated across a school year. Expectations, cultural responsiveness, trust, clarity, and differentiation reinforce one another, so a teacher strengthening any one of these areas strengthens the others at the same time. A student who has felt every one of these practices in the same classroom experiences math not as a subject to survive but as a space where their full effort and full identity both belong. Chapter 10 asks teachers to choose one practice from this chapter to build first, trusting that consistent effort in one area creates room for the rest to follow.

(a) Small, consistent daily choices as the real mechanism of classroom culture change;

(b) the five practices from this chapter reinforcing one another rather than operating in isolation;

(c) one starting practice, chosen with intention, rather than an attempt to change everything at once.

Research Foundations Cited in This Chapter

Gay, G. (2000). *Culturally responsive teaching: Theory, research, and practice*. Teachers College Press.

Hamre, B. K., & Pianta, R. C. (2001). Early teacher-child relationships and the trajectory of children's school outcomes through eighth grade. *Child Development*, *72*(2), 625–638.

Ladson-Billings, G. (1994). *The dreamkeepers: Successful teachers of African American children*. Jossey-Bass.

Ladson-Billings, G. (1995). Toward a theory of culturally relevant pedagogy. *American Educational Research Journal*, *32*(3), 465–491.

Rosenthal, R., & Jacobson, L. (1968). *Pygmalion in the classroom: Teacher expectation and pupils' intellectual development*. Holt, Rinehart & Winston.

Tomlinson, C. A. (2001). *How to differentiate instruction in mixed-ability classrooms* (2nd ed.). Association for Supervision and Curriculum Development.

Supplemental Features for This Chapter

Math Mindset Moment: a short boxed feature challenges teachers to write one sentence of high expectation and one sentence of specific support for a student who is struggling at the moment, then deliver both sentences to that student within the week. The feature closes with the line: “Belief without a plan is just a wish. Pair the two.”

Student Success Story: a composite profile follows a student who had been quiet and disengaged for two years until a new teacher learned to pronounce their name correctly, referenced their community in a word problem, and paired every high expectation with a specific offer of support. The story traces the specific semester when that student’s participation and grades both shifted, tied to the relationship the teacher built before asking for the effort.

Reflection Questions:

(a) identify one student in your classroom who would benefit from a clearer combination of high expectation and specific support;

(b) describe one way you could bring a student’s cultural background or community into an upcoming math lesson;

(c) name one small, repeated action you could take this month to build trust with a specific student or class;

(d) review one recent lesson and identify one point where instructional clarity could have reduced confusion;

(e) design one tiered version of an upcoming task, offering at least two entry points into the same concept.

Chapter Takeaway

A math classroom’s culture traces back to specific, repeatable teacher choices, not to a fixed atmosphere a room either has or lacks (Gay, 2000; Hamre & Pianta, 2001; Ladson-Billings, 1995). High expectations, paired with support, cultural responsiveness, trust, clarity, and differentiation, each remove a barrier to a student's full effort (Rosenthal & Jacobson, 1968; Tomlinson, 2001). A teacher cannot build this culture alone across every hour of a student's day, and this chapter does not ask them to. Families, mentors, and the community around a student extend the same work this chapter has described, carrying it into hours a classroom cannot reach. Chapter 11 turns to that extended circle of support, showing families and mentors how to reinforce the same confidence a strong classroom has already begun to build.

Chapter 11: Families, Mentors, and Math Success

Building a Circle of Support

Getting Started

Success in mathematics is rarely achieved alone. Behind every successful student is a network of people who provide encouragement, guidance, and support. Parents, grandparents, older siblings, coaches, mentors, tutors, and community leaders all play important roles in helping students believe they can succeed, even without knowing every math formula themselves; their encouragement, accountability, and belief in a student's potential often make the greatest difference.

Families use mathematics every day, often without realizing it. Parents compare grocery prices, calculate household budgets, estimate utility bills, and determine how much money to save for vacations or college. A family planning a weekend trip might calculate fuel costs, hotel expenses, meal budgets, and travel time before leaving home. Sharing these experiences with children helps them understand that mathematics is not just a school subject: it is an essential life skill. Conversations like these help remove fear and replace it with curiosity and confidence.

Mentors also provide valuable perspective. A business owner can explain how mathematics helps manage inventory and calculate profits. An engineer can demonstrate how algebra and geometry shape buildings and bridges. A nurse can explain how dosage calculations protect patients. Every caring adult has the opportunity to show young people that mathematics opens doors rather than closes them, and families and mentors, working together, create an environment where students feel supported, challenged, and inspired to pursue excellence.

Real-Life Math Connection

Road Trip

Solution

Distance

420 miles

Speed

60 mph

Time

Distance ÷ Speed

420 ÷ 60

= 7 hours

Fuel

420 miles

÷

28 mpg

=

15 gallons

Gas

15 × $3.40

=

$51

Families use these calculations every day.

Chapter Purpose and Guiding Question

How can families and mentors support a student's mathematical growth without providing direct math instruction?

Chapter 11 addresses the adults around a student outside the classroom: parents, guardians, and mentors who want to help but may not feel qualified to open a math textbook themselves. A family's involvement in a student's education correlated with stronger achievement across dozens of studies, an effect that held regardless of a family's income or education level, provided the involvement connected to what a student was learning in class (Henderson & Mapp, 2002). A parent's own math anxiety can pass to a child through the very act of helping with homework, an effect documented among parents who reported frequent math-anxious help with assignments (Maloney et al., 2015). That finding does not mean a parent who struggled with math should step back; it means that parent needs a different way to help than direct instruction. Chapter 11 supplies that different way: encouragement strategies that work regardless of a parent's own math background, homework support systems that build independence instead of dependence, and community mentoring that extends support beyond what any one household can offer alone. Positive reinforcement, delivered with care about which qualities get praised, builds a student's relationship with math long after a single homework session ends (Mueller & Dweck, 1998). A student's circle of support determines how much of the classroom work from Chapters 1 through 10 holds once that student walks out the door. The remaining sections provide families and mentors with specific, low-barrier tools to strengthen that circle.

Learning Objectives

Chapter 11 sets five learning outcomes for families and mentors to reach before its final page:

(a) explain why family involvement supports achievement even when a parent has limited math background (Henderson & Mapp, 2002);

(b) identify one way to help with math homework that avoids passing on math anxiety (Maloney et al., 2015);

(c) practice giving process-focused praise instead of praise aimed at fixed ability (Mueller & Dweck, 1998);

(d) set up one homework support routine that builds a student's independence rather than dependence;

(e) identify one community mentor or program that could extend support beyond the household (DuBois et al., 2011).

Section Outline

I. Why Family Involvement Matters, Even Without Math Expertise

A parent does not need to know how to factor a quadratic to help a child succeed in math; a parent needs to show, through consistent attention, that math matters in that household. Family involvement has been linked to student achievement across a wide range of studies, and the strongest results were observed when families connected their involvement to specific learning goals rather than to general encouragement alone (Henderson & Mapp, 2002). Asking a specific question, such as "what is one thing you learned in math today," communicates investment without requiring a parent to solve a single problem. Involvement, defined this way, belongs to every family regardless of what math instruction looked like when that family's adults were in school.

(a) Family involvement linked to achievement across income and education levels, when tied to specific learning goals (Henderson & Mapp, 2002);

(b) specific questions that show investment without requiring math expertise;

(c) involvement redefined as attention and structure, not as an ability to solve the homework.

II. The Anxiety Trap: How a Parent's Math Story Can Become a Child's

A parent's own difficult history with math does not stay contained within that parent; it can be passed on to a child through tone, body language, and the specific way a parent responds when homework gets hard. Children of math-anxious parents learned less math over the school year and developed more math anxiety themselves, but only when those parents provided frequent help with homework, a pattern that pointed to the help itself as the transmission mechanism (Maloney et al., 2015). A parent who says "I was never good at this either" while sighing over a worksheet passes along more than a fact; that parent passes along a script for how to feel about math. Breaking this pattern does not require a parent to fake confidence; it requires a parent to hand off direct help to another resource, such as a tutor, a sibling, or the strategies in this chapter, while still showing up with encouragement.

(a) The specific mechanism of transmission: frequent, anxious help with homework (Maloney et al., 2015);

(b) parent language and body language that can communicate fear of math without anyone intending it;

(c) handing off direct instructional help while still providing encouragement and structure.

III. Homework Support Without Taking Over

Support and rescue look similar from a distance but produce opposite results: rescue removes the struggle a student needs, while support stays present without removing it. A parent who sits beside a student, asks questions instead of supplying answers, and tolerates a visible struggle offers support; a parent who takes the pencil and finishes the problem offers rescue. A consistent homework routine, including a fixed time, a quiet space, and a clear starting ritual, does more to support achievement than any single moment of direct help ever could. Community and mentoring resources, including tutoring programs and structured mentoring relationships, provided families with another option when a specific concept required expertise that no one in the household possessed (DuBois et al., 2011).

(a) Support defined as presence and questions, not answers handed over;

(b) a consistent homework routine: fixed time, quiet space, clear starting ritual;

(c) outside resources, including tutoring and mentoring, used for concepts beyond a family's own expertise (DuBois et al., 2011).

IV. Positive Reinforcement: Praising the Process, Not the Person

Not all praise builds confidence, and some praise, offered with the best intentions, undermines the very confidence it was meant to build. Students praised for intelligence after a success cared more about looking smart than about learning, and after a later failure, those same students showed less persistence and worse performance than students praised for their effort and strategy (Mueller & Dweck, 1998). "You worked hard on that" builds a different relationship with math than "you're so smart," since the first statement points to something a student controls and can repeat. A family that shifts its praise toward process gives a student the same growth-oriented message the classroom has been building since Chapter 2.

(a) The measured difference in outcomes between praise for intelligence and praise for effort (Mueller & Dweck, 1998);

(b) specific praise language that names a strategy or a behavior, not a fixed trait;

(c) consistency between the praise a student hears at home and the praise a student hears at school.

V. Community Mentoring: Extending the Circle Beyond the Household

A single household cannot provide all the support a student needs, and mentoring programs exist to fill that gap by providing a caring, consistent adult outside the immediate family. Structured mentoring relationships produced meaningful gains across academic, social, and emotional outcomes for young people, with the strongest effects appearing in programs that offered consistent contact over an extended period (DuBois et al., 2011). A mentor who shares a student's background, including a mentor who once struggled with math and found a way through it, offers a model of possibility a parent cannot always provide alone. Community mentoring works alongside family involvement rather than replacing it, adding another consistent adult to the circle surrounding a student.

(a) Mentoring's measured academic, social, and emotional benefits for young people (DuBois et al., 2011);

(b) consistent, extended contact as a predictor of a mentoring relationship's effectiveness;

(c) mentors who share a student's background as a specific source of representation and possibility.

VI. Building a Family Culture of Math Confidence

A family culture around math forms over time through hundreds of small moments rather than a single dramatic conversation about the subject's importance. Families who folded math into daily life, through cooking, budgeting, sports statistics, or games, gave students repeated proof that math exists outside a textbook and belongs to ordinary life. A family that talks about its own current learning, including a parent picking up a new skill as an adult, models the same growth mindset the classroom has spent ten chapters building. Chapter 11 closes with an invitation for every family to choose one strategy from this chapter and practice it for a single month, trusting that a small, consistent change in the household can shift a student's entire relationship with math.

(a) Math folded into daily routines: cooking, budgeting, sports statistics, games;

(b) a parent modeling their own current learning as an adult;

(c) one strategy chosen and practiced for a full month, rather than an attempt to change everything at once.

Research Foundations Cited in This Chapter

DuBois, D. L., Portillo, N., Rhodes, J. E., Silverthorn, N., & Valentine, J. C. (2011). How effective are mentoring programs for youth? A systematic assessment of the evidence. *Psychological Science in the Public Interest, 12*(2), 57–91.

Henderson, A. T., & Mapp, K. L. (2002). *A new wave of evidence: The impact of school, family, and community connections on student achievement*. Southwest Educational Development Laboratory.

Maloney, E. A., Ramirez, G., Gunderson, E. A., Levine, S. C., & Beilock, S. L. (2015). Intergenerational effects of parents' math anxiety on children's math achievement and anxiety. *Psychological Science*, *26*(9), 1480–1488.

Mueller, C. M., & Dweck, C. S. (1998). Praise for intelligence can undermine children's motivation and performance. *Journal of Personality and Social Psychology*, *75*(1), 33–52.

Supplemental Features for This Chapter

Math Mindset Moment: a short boxed feature invites a parent or guardian to share one honest sentence about their own math history with their student, followed by one sentence of belief in that student's future. The feature closes with the line: "My story is not your story. Your chapter is still being written."

Student Success Story: a composite profile follows a student whose parent had, for years, said "I was never a math person" in front of them, until that parent learned to say "I am learning alongside you" instead and began asking process-focused questions about homework. The story traces the specific school year when the household's language shifted and the student's confidence, then grades, followed close behind.

Reflection Questions:

(a) describe one thing you have said about your own math ability in front of your student, and consider how that message may have landed;

(b) identify one specific question you could ask about math homework that does not require you to solve a problem;

(c) describe your current homework routine and identify one change that could turn rescue into support;

(d) practice giving process-focused praise this week and note how your student responds compared with praise for being smart;

(e) identify one community resource, tutor, or mentor that could extend support beyond what your household offers alone.

Chapter Takeaway

Family involvement supports achievement, and it requires attention and structure rather than math expertise (Henderson & Mapp, 2002). A parent's own math anxiety can pass to a child through the specific act of anxious help, which means the solution involves handing off direct instruction rather than withdrawing from the relationship altogether (Maloney et al., 2015). Praise aimed at effort and process built a stronger relationship with math than praise aimed at fixed intelligence, a small language shift every family can practice starting today (Mueller & Dweck, 1998). Community mentoring extends this circle of support beyond any single household, adding consistent adults who can offer what a family alone cannot (DuBois et al., 2011). Chapter 12 turns from this circle of support to the daily habits a student builds inside it, showing how consistency and routine turn every strategy in this book into lasting, disciplined practice.

Chapter 12: The Mathematics of Discipline

Consistency Creates Confidence

Getting Started

Discipline is one of the greatest predictors of success in mathematics. While talent may give someone a head start, discipline determines how far they will go. Successful math students understand that consistent effort, daily practice, and responsible study habits produce better results than last-minute cramming. Discipline means choosing long-term success over short-term comfort, even when motivation is low.

Think about someone training for a marathon: no one wakes up one morning and decides to run 26.2 miles without preparation, but instead follows a training schedule, gradually increasing their distance each week. Mathematics requires the same steady commitment. Imagine saving money to purchase your first car. If you save $25 each week, you can calculate how much you will have after six months or a year. Small, consistent actions eventually produce significant results. Likewise, solving just five additional math problems each day may seem insignificant at first, but over an entire school year, that practice adds up to thousands of opportunities to strengthen your skills.

Develop routines that support your success. Set aside a specific time each day to review notes, complete homework, and practice challenging concepts. Turn off distractions, stay organized, and remain committed even when lessons become difficult. Discipline transforms ordinary students into extraordinary achievers because it teaches them that success is built one decision, one assignment, and one day at a time.

Real-Life Math Connection

Save $40 each week.

Solution

After one year

40 × 52

=

$2,080

Small consistent habits produce remarkable long-term results.

Chapter Purpose and Guiding Question

How does consistent practice, more than talent, determine a student's progress in mathematics?

Chapter 12 argues that consistency, not talent, accounts for most of the difference between a student who improves in math and one who stays stuck. Discipline sounds like a character trait some students possess, and others lack, but the research covered in this chapter treats discipline as a set of specific, learnable systems instead. Practice sessions spread across multiple days produced stronger long-term retention than the same amount of practice crammed into a single session, a finding that held across a wide range of subjects and skill types (Cepeda et al., 2006). Expert performers across fields accumulated their skill through structured, effortful practice sustained over years, not through raw talent alone, which means a student's current skill level says less about their ceiling than their practice system does (Ericsson et al., 1993). Specific, challenging goals produced higher performance than vague goals such as "do your best," provided the goal came paired with a clear plan for reaching it (Locke & Latham, 2002). Chapter 12 turns these findings into four concrete systems: daily practice routines, goal-setting structures, study schedules, and accountability partnerships. Self-regulated learners cycled through planning, monitoring, and reflecting on their own work, a cycle any student can build with the right structure and enough repetition (Zimmerman, 2002). The remaining sections give students that structure, one system at a time.

Learning Objectives

Chapter 12 sets five learning outcomes for students to reach before its final page:

(a) explain why spaced practice produced stronger retention than cramming the same material into one session (Cepeda et al., 2006);

(b) describe the role of structured, effortful practice in building expert-level skill over time (Ericsson et al., 1993);

(c) write one specific, challenging math goal paired with a clear plan for reaching it (Locke & Latham, 2002);

(d) build a weekly study schedule that includes fixed times for math practice;

(e) identify one accountability partner and describe how that partnership will support consistency.

Section Outline

I. Discipline as a Skill, Not a Personality Trait

Discipline gets treated like a trait a person is born with, present in some students and absent in others, when the research points to something more useful: discipline as a set of systems a person builds and refines. A student who seems to have discipline built in has, in most cases, built invisible routines over years, not been born immune to the pull of distraction. Removing the mystery around discipline gives every student permission to build the same systems, regardless of how consistent their habits have been so far. Chapter 12 treats every strategy that follows as learnable, not as a test of character a student has already passed or failed.

(a) Discipline reframed as a built system rather than an inborn trait;

(b) the invisible routines behind students who appear consistent without effort;

(c) every strategy in this chapter treated as learnable, regardless of a student's current habits.

II. Daily Math Practice: Why Spacing Beats Cramming

A single long study session before a test feels productive in the moment but produces weaker long-term retention than the same amount of time spread across several shorter sessions. Practice distributed across days, rather than massed into one sitting, produced stronger recall on a later test across a wide range of subjects and skill types, an effect strong enough to hold up in nearly 200 experiments (Cepeda et al., 2006). A short daily math practice, even ten or fifteen minutes, builds and maintains a skill with more staying power than a single ninety-minute cram session the night before a test. Chapter 4's organizational systems and Chapter 5's note-taking habits give a student the materials; daily spaced practice gives those materials a reason to stay in memory.

(a) The distributed practice effect: spaced sessions outperforming a single massed session (Cepeda et al., 2006);

(b) a short daily practice routine, ten to fifteen minutes, as a realistic minimum;

(c) daily practice connected to the organizational and note-taking systems built in Chapters 4 and 5.

III. Deliberate Practice: Practice With a Purpose

Not all practice builds skill at the same rate; practice aimed at a specific weakness, performed with full attention and paired with feedback, builds skill far faster than repetition alone. Expert performers across music, sports, and academic fields accumulated their skill through years of structured practice targeted at the exact edge of their current ability, not through unfocused repetition of what they already knew how to do (Ericsson et al., 1993). A student practicing with intention chooses the specific problem type that gave them trouble last week, rather than repeating twenty problems they can already solve with ease. Deliberate practice asks more of a student than passive repetition, but it also returns far more growth for the same amount of time invested.

(a) Deliberate practice aimed at a specific weakness, not general repetition;

(b) the link between structured, effortful practice and expert-level skill development (Ericsson et al., 1993);

(c) choosing practice problems based on last week's specific mistakes, not comfort or ease.

IV. Goal-Setting Systems That Drive Behavior

A vague goal, such as "get better at math," gives a student nothing to act on tomorrow morning; a specific goal gives a student a target that shapes behavior starting today. Specific, challenging goals produced higher performance across studies than vague or easy goals, provided the goal setter also received feedback on progress toward the goal (Locke & Latham, 2002). A goal such as "solve eight of ten practice problems without error before Friday's quiz" gives a student both a target and a way to know whether that target has been met. Chapter 12 asks students to write goals with three parts: a specific number, a clear deadline, and a plan for the practice sessions that will get them there.

(a) Specific, challenging goals compared with vague goals such as "do your best" (Locke & Latham, 2002);

(b) feedback on progress as a required companion to any goal, not an optional extra;

(c) a three-part goal format: a specific number, a clear deadline, a practice plan.

V. Study Schedules: Turning Intention Into a Plan

Intention alone seldom survives a busy week; a student who plans to "practice math sometime" loses to a student who has already claimed a specific block of time on a specific day. A written study schedule turns a vague intention into a concrete plan, assigning practice sessions to fixed days and times the way an appointment claims a place on a calendar. Self-regulated learners planned their study time in advance, monitored whether they followed that plan, and adjusted it based on what worked, cycling through the same three steps over the course of a term (Zimmerman, 2002). A schedule does not need to be complicated to work; it needs to be specific enough that a student can look at it and know without guessing what to do next.

(a) A written schedule assigning practice to specific days and times, not vague intention;

(b) the planning, monitoring, and adjusting cycle of self-regulated learning (Zimmerman, 2002);

(c) a schedule specific enough to remove any guesswork about what to do next.

VI. Accountability Partners: Consistency Through Connection

A plan known only to the student who made it is easier to abandon than a plan another person is watching, since private failure carries a smaller cost than failure someone else has already noticed. An accountability partner, whether a classmate, a family member, or a mentor, gives a student a specific person to report progress to, adding a social cost to skipping a planned practice session. A brief daily or weekly check-in, as simple as a text message confirming whether a scheduled practice took place, keeps a system alive during the weeks when motivation alone would not have been enough. Chapter 12 closes with a direct connection between every system in this chapter and the circle of support built in Chapter 11, since a goal, a schedule, and a practice routine all work better inside a relationship than in isolation.

(a) An accountability partner: a classmate, family member, or mentor tracking progress alongside the student;

(b) brief, regular check-ins that report whether a planned practice session happened;

(c) every system in this chapter working through the circle of support introduced in Chapter 11.

Research Foundations Cited in This Chapter

Cepeda, N. J., Pashler, H., Vul, E., Wixted, J. T., & Rohrer, D. (2006). Distributed practice in verbal recall tasks: A review and quantitative synthesis. *Psychological Bulletin*, *132*(3), 354–380.

Ericsson, K. A., Krampe, R. T., & Tesch-Römer, C. (1993). The role of deliberate practice in the acquisition of expert performance. *Psychological Review*, *100*(3), 363–406.

Locke, E. A., & Latham, G. P. (2002). Building a practically useful theory of goal setting and task motivation: A 35-year odyssey. *American Psychologist*, *57*(9), 705–717.

Zimmerman, B. J. (2002). Becoming a self-regulated learner: An overview. *Theory Into Practice*, *41*(2), 64–70.

Supplemental Features for This Chapter

Math Mindset Moment: a short, boxed feature asks students to schedule their next three math practice sessions on a calendar before finishing the chapter, treating each session as a fixed appointment. The feature closes with the line: “Discipline is not a feeling. Discipline is a calendar entry I keep.”

Student Success Story: a composite profile follows a student who studied in long, irregular bursts before every test and forgot most of it within a month, until a fixed fifteen-minute daily schedule and a classmate accountability partner replaced the old pattern. The story traces the specific grading period during which consistent, spaced practice yielded a stronger result than any single long study session had ever produced.

Reflection Questions:

(a) describe your current approach to math practice: consistent and spaced, or crammed and irregular;

(b) write one specific, challenging math goal for this month, including a number and a deadline;

(c) build a study schedule for next week, assigning math practice to specific days and times;

(d) identify one accountability partner and describe the check-in system you will use together;

(e) choose one specific weakness from a recent assignment and design a short deliberate-practice session targeting it.

Chapter Takeaway

Consistency, not raw talent, accounts for most of the gap between a student on a steady path of improvement and one who stays stuck (Ericsson et al., 1993). Spaced daily practice, deliberate focus on specific weaknesses, written goals, a real schedule, and an accountability partner each turn good intentions into a system that runs even on a low-motivation day (Cepeda et al., 2006; Locke & Latham, 2002; Zimmerman, 2002). None of these systems require a personality change; each one requires a structure a student can build this week and keep running. A student who has read this far now has almost every tool this book offers: mindset, vocabulary, organization, writing, reasoning, focus, error analysis, identity, teacher and family support, and now discipline. Chapter 13 turns to the question every one of these tools serves: What does all this practice make possible in a student's future?

Chapter 13: Real-Life Math and Future Opportunities

Connecting Mathematics to Purpose

Getting Started

Many students ask, “When will I ever use this?” The answer is simple: every day. Mathematics is woven into nearly every career and every important financial decision we make. Whether you dream of becoming a pilot, entrepreneur, teacher, electrician, chef, architect, mechanic, computer programmer, nurse, or professional athlete, mathematics provides the foundation for success. Learning math today prepares you to solve the challenges of tomorrow.

Consider someone interested in owning a small business. Before opening the doors, they must calculate startup costs, estimate monthly expenses, determine product prices, predict profits, and manage taxes. A contractor uses geometry to estimate the amount of flooring materials needed. A nurse calculates medication dosages with precision. A pilot uses speed, distance, fuel consumption, and weather data to safely complete every flight. Even social media influencers analyze engagement statistics and advertising revenue to grow their businesses. Mathematics empowers people to make informed decisions with confidence.

The future belongs to problem-solvers. Employers consistently seek individuals who can reason with clarity, analyze data, and make sound decisions. Mathematics teaches these valuable habits of mind. Rather than asking, “Will I ever use this?” begin asking, “How can this knowledge prepare me for my future?” Every equation you solve today strengthens the skills that will help shape your career, your finances, and your opportunities for years to come.

Real-Life Math Connection

A small business buys products for $8 each.

Solution

Selling Price

$15

Profit

15 − 8

=

$7 each

Selling 500 items

500 × 7

=

$3,500 profit

Entrepreneurs depend on mathematics every day.

Chapter Purpose and Guiding Question

How does connecting mathematics to real-life goals and careers change a student's motivation to persist?

Chapter 13 answers a question students have asked in some form since Chapter 1: When will I ever use this? A student who cannot connect a math skill to any future goal has little reason to push through the hard moments this book has spent twelve chapters teaching them to survive. Writing about how course material connected to their own lives raised both interest and grades for students who expected to struggle, an effect strong enough to move a letter grade for the students who needed it most (Hulleman & Harackiewicz, 2009). Expectancy and value combined to predict whether a student engaged with a subject, meaning a student needs to believe both that they can succeed and that success matters for something they care about (Eccles & Wigfield, 2002). Chapter 13 supplies that second half of the equation: concrete, specific connections between math and the careers, goals, and daily decisions a student already cares about. STEM fields, financial decisions, entrepreneurship, data and technology, and the trades all depend on math in ways a classroom worksheet seldom makes visible. Structured pathways into STEM fields produced strong outcomes for students when those pathways paired rigorous math preparation with a clear sense of where that preparation led (Maton et al., 2000). The remaining sections walk through five specific domains, giving students language for exactly what math builds toward in each one.

Learning Objectives

Chapter 13 sets five learning outcomes for students to reach before its final page:

- (a) explain why connecting math to a personal goal increases motivation to persist through a hard problem (Hulleman & Harackiewicz, 2009);
- (b) identify one STEM career that depends on math skills taught in this book (Maton et al., 2000);
- (c) apply one math concept to a financial literacy scenario, such as budgeting or interest (Lusardi & Mitchell, 2014);
- (d) describe one way an entrepreneur might use math to start or run a business;
- (e) identify one trade or engineering career and describe the specific math skills that career requires.

Section Outline

I. Why Relevance Changes Motivation

Motivation responds to more than confidence; a student can believe they are capable of learning math and still disengage if that math never connects to anything the student wants for themselves. A brief writing exercise asking students to explain how course material related to their own lives increased interest and grades over an entire semester, with the largest gains among students who most doubted their own ability (Hulleman & Harackiewicz, 2009). Expectancy for success and the value placed on a task combine to predict engagement, meaning relevance functions as one half of the motivation equation that confidence-building strategies alone cannot supply (Eccles & Wigfield, 2002). Chapter 13 treats relevance as a strategy in its own right, not a nice bonus layered on top of the real work of building skill.

(a) A brief relevance-writing exercise linked to measurable gains in interest and grades (Hulleman & Harackiewicz, 2009);

(b) expectancy and value as two separate ingredients in student motivation (Eccles & Wigfield, 2002);

(c) relevance treated as a core strategy rather than an optional addition to skill-building.

II. STEM Careers: Where Math Opens the Door

Science, technology, engineering, and math careers share a common gatekeeper: the math coursework a student completes long before applying for any job in the field. Structured programs pairing rigorous math and science preparation with a strong peer and mentor community produced high rates of advanced degree completion among students entering STEM fields, well above typical national rates for that group (Maton et al., 2000). A student does not need to already know they want to become an engineer to benefit from strong math preparation; that preparation keeps every STEM door open for a decision the student can make later. Naming specific STEM careers, from data scientists to civil engineers to pharmacists, gives an abstract subject a concrete destination.

(a) Math coursework functioning as a gatekeeper for entry into STEM majors and careers;

(b) structured peer and mentor communities linked to strong STEM outcomes for participating students (Maton et al., 2000);

(c) specific STEM career examples named and connected to specific math skills.

III. Financial Literacy: Math as a Life Skill

Every adult manages a budget, a loan, an interest rate, or a paycheck, which means financial literacy affects every student's future, regardless of the career they eventually choose. Financial knowledge served as a form of human capital, and people with higher financial literacy made measurably better decisions about saving, borrowing, and retirement throughout their lives (Lusardi & Mitchell, 2014). A single lesson connecting percentages to credit card interest or linear equations to a monthly budget turns an abstract skill into a tool a student will use the week they get their first paycheck. Financial literacy gives math a stake in a student's life that has nothing to do with a grade on a report card.

(a) Financial knowledge treated as a form of human capital with measurable economic returns (Lusardi & Mitchell, 2014);

(b) specific math-to-finance connections: percentages and interest, equations and budgets;

(c) financial literacy as a math application every student will use, regardless of career.

IV. Entrepreneurship: Math as a Tool for Building Something New

An entrepreneur uses math constantly, calculating costs, prices, profit margins, and break-even points long before ever writing a business plan. A student who wants to start a business, sell a product, or run a service needs the same proportional reasoning and algebraic thinking this book has spent thirteen chapters building, applied to a spreadsheet instead of a worksheet. Framing math as the language of business ownership gives students who see themselves as future entrepreneurs a direct, personal reason to master the material in front of them. Chapter 13 encourages students to design a simple business idea and calculate its basic costs and potential profit to put this connection into practice.

(a) The specific math behind entrepreneurship: cost, price, margin, break-even analysis;

(b) proportional reasoning and algebra applied to a business spreadsheet rather than a worksheet;

(c) a simple student business idea used as a hands-on math application.

V. Data and Technology: The Math Behind the Modern World

Every app, algorithm, and piece of technology a student uses daily runs on math, even when the math stays hidden behind a clean interface. Careers in data science, software engineering, and artificial intelligence depend on statistics, algebra, and logical reasoning as core, everyday tools rather than occasional requirements. A student who enjoys technology but dislikes math often has not yet seen the connection between the two, since the interesting part of technology work happens through the math rather than around it. Naming this connection in plain terms gives a tech-curious student a reason to treat math class as training for a career that already falls within their range of interests.

(a) Statistics, algebra, and logical reasoning as core tools in data science and software careers;

(b) the math hidden behind everyday technology and interfaces;

(c) technology interest reframed as a direct reason to invest in math class.

VI. Trades and Engineering: Math You Can Build With Your Hands

Trades including construction, electrical work, plumbing, and welding depend on measurement, geometry, and proportional reasoning applied under real physical constraints, not abstract numbers on a page. A carpenter calculating an angle, an electrician calculating a load, and an engineer calculating a stress tolerance all use the same underlying math skills this book has spent thirteen chapters building, applied toward a physical result a person can see and touch. Careers like these offer a path to stable, well-paid work that does not require a four-year degree, provided a student has first built solid math skills. Chapter 13 closes with an invitation for students to research one trade or engineering career, tracing the specific math skills that career depends on, from the classroom straight through to the job site.

(a) Measurement, geometry, and proportional reasoning applied under real physical constraints;

(b) specific trade examples: carpentry, electrical work, plumbing, welding;

(c) a stable career path that rewards strong math skills without requiring a four-year degree.

Research Foundations Cited in This Chapter

Eccles, J. S., & Wigfield, A. (2002). Motivational beliefs, values, and goals. *Annual Review of Psychology*, *53*, 109–132.

Hulleman, C. S., & Harackiewicz, J. M. (2009). Promoting interest and performance in high school science classes. *Science*, *326*(5958), 1410–1412.

Lusardi, A., & Mitchell, O. S. (2014). The economic importance of financial literacy: Theory and evidence. *Journal of Economic Literature*, *52*(1), 5–44.

Maton, K. I., Hrabowski, F. A., III, & Schmitt, C. L. (2000). African American college students excelling in the sciences: College and postcollege outcomes in the Meyerhoff Scholars Program. *Journal of Research in Science Teaching*, *37*(7), 629–654.

Supplemental Features for This Chapter

Math Mindset Moment: a short, boxed feature asks students to write three sentences that connect today's math lesson to a career or goal they care about, even if the connection feels like a stretch at first. The feature closes with the line: “Not just a lesson. A tool I am building for my own future.”

Student Success Story: a composite profile follows a student who saw no point in algebra until a unit connecting linear equations to starting a small lawn-care business gave the subject a concrete purpose. The story traces the semester when that student's grades and engagement both rose, tied to a single relevance-writing assignment that required them to name their own goal in their own words.

Reflection Questions:

(a) name one career, trade, or business idea you are curious about and describe the math skills it likely requires;

(b) write three sentences connecting a recent math lesson to your own life or future goals;

(c) calculate a simple budget or interest scenario using a math skill from this book;

(d) research one STEM, trade, or engineering career and describe the path from high school math to that career;

(e) design one simple business idea and calculate its basic costs and potential profit.

Chapter Takeaway

Confidence is not something people are born with; it is something they build through preparation, persistence, and experience. Every successful mathematician once struggled with difficult concepts. The difference is that they refused to let temporary setbacks define their future. Confidence grows each time you solve a problem independently, ask a thoughtful question, or continue working when a solution is not immediately obvious.

Think about learning to ride a bicycle. The first attempts are often filled with uncertainty and occasional falls. Yet each attempt builds better balance, coordination, and courage. Mathematics follows the same pattern. Imagine calculating the monthly payment for a new bicycle using a payment plan. At first, the calculations involving interest rates and monthly installments may appear complicated.

However, as you practice similar problems, they become easier and more familiar. Each success builds confidence for the next challenge.

Confident math students are not fearless; they simply refuse to allow fear to control their decisions, preparing consistently, asking for help when needed, celebrating progress, and embracing challenges as opportunities to learn. Every problem solved strengthens their belief that they are capable of achieving even greater success. Confidence grows one solution, one lesson, and one victory at a time.

Relevance functions as its own motivational strategy, distinct from confidence and separate from raw skill (Eccles & Wigfield, 2002; Hulleman & Harackiewicz, 2009). STEM careers, financial decisions, entrepreneurship, technology, and the trades each depend on math in specific, nameable ways, giving a student concrete answers to the question of why this subject matters (Lusardi & Mitchell, 2014; Maton et al., 2000). A student who can name their own connection to math holds a reason to keep practicing that no worksheet alone could ever supply. Every strategy from earlier chapters (mindset, vocabulary, organization, focus, and discipline) now has a destination attached to it. Chapter 14 turns to the final piece of the puzzle: building the specific, practiced confidence a student needs to walk into a test, a classroom, or an opportunity and perform at the level all this preparation has made possible.

Chapter 14: Building the Confident Math Student

From Fearful to Fearless

Getting Started

Confidence is not something people are born with; it is something they build through preparation, persistence, and experience. Every successful mathematician once struggled with difficult concepts. The difference is that they refused to let temporary setbacks define their future. Confidence grows each time you solve a problem independently, ask a thoughtful question, or continue working when a solution is not immediately obvious.

Think about learning to ride a bicycle. The first attempts are often filled with uncertainty and occasional falls. Yet each attempt builds better balance, coordination, and courage. Mathematics follows the same pattern. Imagine calculating the monthly payment for a new bicycle using a payment plan. At first, the calculations involving interest rates and monthly installments may appear complicated. However, as you practice similar problems, they become easier and more familiar. Each success builds confidence for the next challenge.

Confident math students are not fearless; they simply refuse to allow fear to control their decisions. They prepare consistently, ask for help when needed, celebrate progress, and embrace challenges as opportunities to learn. Every problem solved strengthens their belief that they are capable of achieving even greater success. Confidence grows one solution, one lesson, and one victory at a time.

Real-Life Math Connection

Buying a Bicycle

Solution

Cost

$900

Down Payment

$150

Remaining Balance

$750

12 Monthly Payments

750 ÷ 12

=

$62.50 per month

Confidence grows as students learn to solve real financial problems independently.

Chapter Purpose and Guiding Question

What specific practices allow a student to walk into a test or a moment of real difficulty with confidence rather than dread?

Chapter 14 gathers every strategy this book has built: mindset, vocabulary, organization, reasoning, focus, identity, support, and discipline, and asks a single question: what does a confident math student look like on test day, in class, and in a moment of real difficulty? Confidence is not a personality a student either has or lacks; it is a belief in one's own capability, built through repeated evidence of success and through specific practices a student can rehearse on purpose (Bandura, 1977). A short writing exercise asking students to affirm their own values before a stressful academic moment reduced the achievement gap tied to stereotype threat and produced grade gains that lasted for years afterward (Cohen et al., 2006). A separate but related exercise, asking students to write about their worries just before an exam, freed up mental resources that anxiety would otherwise consume, raising exam scores among the students who felt the most test anxiety to begin with (Ramirez & Beilock, 2011). Stereotype threat, first identified as a specific, situational pressure rather than a fixed trait, can undermine performance under the very testing conditions in which a student most needs to perform well (Steele & Aronson, 1995). Chapter 14 turns these findings into four rehearsed practices: self-affirmation before a stressful task, expressive writing before a test, a fearless-response routine for handling in-the-moment difficulties, and a specific plan for walking into any test with belief rather than dread. A student who has practiced growth-oriented self-talk since Chapter 2 already holds half of this chapter's foundation (Dweck, 2006). The remaining sections turn confidence from an abstract goal into a specific set of moves a student can rehearse the same way they rehearse a math procedure.

Learning Objectives

Chapter 14 sets five learning outcomes for students to reach before its final page:

(a) explain how repeated evidence of capability builds durable self-efficacy over time (Bandura, 1977);

(b) practice a brief self-affirmation exercise before a stressful academic task (Cohen, Garcia, Apfel, & Master, 2006);

(c) use expressive writing to reduce anxiety before a test or quiz (Ramirez & Beilock, 2011);

(d) describe stereotype threat as a situational pressure rather than a fixed personal limitation (Steele & Aronson, 1995);

(e) build a personal fearless-response routine for the moment a hard problem first appears on the page.

Section Outline

I. What Confidence Is

Confidence gets treated as a mood, present on a good day and absent on a bad one, when research points to something more stable underneath the mood: a belief in one's own capability built through evidence, not through hope alone. Expectations of personal capability shape whether a person starts a difficult task, how much effort they spend on it, and how long they continue once the task becomes difficult (Bandura, 1977). A student does not need to feel confident before starting a hard problem; a student needs enough evidence of past success to start the problem anyway, and confidence tends to follow action rather than precede it. Chapter 14 treats confidence as a practiced skill built through specific exercises, not a trait a student is waiting to discover.

(a) Confidence defined as a belief built through evidence, not a mood that arrives on its own;

(b) the four sources of self-efficacy: past success, watching others succeed, encouragement, and one's own physical state (Bandura, 1977);

(c) action taken first, with confidence following close behind rather than leading the way.

II. Self-Affirmation: A Brief Exercise With a Lasting Effect

A short writing exercise, asking a student to name a personal value and explain why it matters to them, sounds too small to matter in a math classroom, yet the evidence says otherwise. Students who completed a brief values-affirmation writing exercise before a stressful academic term closed a measurable share of the achievement gap tied to stereotype threat, with the strongest effects appearing among students who entered the term most at risk (Cohen et al., 2006). The exercise works because it reminds a student of a stable sense of self-worth that does not depend on the outcome of any single test, freeing that student to take an academic risk without feeling that their entire identity rides on the result. Chapter 14 asks students to complete this exact exercise before their next major math assessment, writing three to five sentences about a value that matters to them and why.

(a) A brief values-affirmation writing exercise completed before a stressful academic term (Cohen, Garcia, Apfel, & Master, 2006);

(b) self-worth grounded in a stable value rather than in the outcome of a single test;

(c) the exercise scheduled before a student's next major math assessment, not saved for a future crisis.

III. Expressive Writing: Emptying the Mind Before a Test

Test anxiety consumes working memory, the exact mental resource a student needs to hold a problem in mind while solving it, which means anxious thoughts and math reasoning compete for the same limited space. Students who spent 10 minutes writing about their worries right before an exam outperformed students who sat in silence, with the largest gains among students who reported the highest levels of test anxiety (Ramirez & Beilock, 2011). The writing does not need to solve the worry; it only needs to move the worry from working memory onto the page, freeing that mental space for the test itself. Chapter 14 asks students to keep one index card and one pencil in their bag on test day, ready for a brief expressive-writing session in the minutes before the test begins.

(a) Test anxiety competing with math reasoning for the same limited working memory;

(b) a ten-minute expressive-writing exercise linked to higher exam scores, with the largest gains for anxious students (Ramirez & Beilock, 2011);

(c) a simple test-day tool: one index card and one pencil, ready before the test begins.

IV. Naming Stereotype Threat So It Loses Its Power

A student walking into a test aware that their group carries a negative stereotype about math ability can feel an added layer of pressure that has nothing to do with actual preparation, a pressure that can lower performance on its own. Stereotype threat operated as a situational pressure tied to a specific testing condition, not as a fixed limitation belonging to any student or group, and performance improved once that pressure eased (Steele & Aronson, 1995).

Naming stereotype threat out loud, on its own, and explaining to students that the extra pressure they may feel is a documented, temporary phenomenon rather than proof of anything about their own ability, has itself been shown to reduce its effect. A student who understands this pressure for what it is can recognize it in the moment and continue working rather than reading it as a verdict on their own capability.

(a) Stereotype threat identified as situational and temporary, not a fixed trait (Steele & Aronson, 1995);

(b) naming the pressure out loud as a step that itself reduces its power;

(c) a student's ability to recognize the pressure in the moment and keep working instead of surrendering to it.

V. The Fearless-Response Routine: What to Do When a Hard Problem Appears

Every student, regardless of preparation, will at some point face a problem that looks unfamiliar the moment it appears on the page, and what that student does in the next ten seconds matters more than any amount of prior practice. A rehearsed routine, practiced until it becomes automatic, replaces panic with a series of concrete next steps: pause, breathe, reread the problem, identify one thing that is known, and try one strategy before deciding the problem is unsolvable. A growth-oriented response to difficulty, treating a hard problem as a chance to build skill rather than a threat to one's identity, changes how a student interprets that same ten seconds of uncertainty (Dweck, 2006). Chapter 14 asks students to write their own four-step fearless-response routine and practice it on a problem chosen because it is hard, before they ever need the routine on a real test.

(a) A rehearsed, automatic routine for the first ten seconds after a hard problem appears;

(b) four concrete steps: pause, breathe, reread, identify one known fact, then attempt one strategy;

(c) difficulty reframed as a skill-building opportunity rather than an identity threat (Dweck, 2006).

VI. Walking Into Any Test With a Plan

A student who walks into a test with a plan carries a different posture than a student who walks in hoping for the best, and that posture shows up in performance as much as in preparation. Chapter 14's full test-day plan combines every practice from this chapter into a single sequence: a values-affirmation sentence written the night before, ten minutes of expressive writing before the test begins, a private reminder that any extra pressure felt in the room is temporary and situational, and the fearless-response routine held ready for the first hard problem on the page. None of these steps replaces the content knowledge built across thirteen earlier chapters; each one protects that knowledge from anxiety that would otherwise block it from showing up on the page. A student who has practiced this plan on smaller quizzes will find it far easier to trust on the day it matters most.

(a) A full test-day sequence combining affirmation, expressive writing, and the fearless-response routine;

(b) confidence practices that protect existing knowledge rather than replace the need to study;

(c) smaller quizzes used as low-stakes practice for the exact plan a student will use on a major exam.

Research Foundations Cited in This Chapter

Bandura, A. (1977). Self-efficacy: Toward a unifying theory of behavioral change. *Psychological Review*, *84*(2), 191–215.

Cohen, G. L., Garcia, J., Apfel, N., & Master, A. (2006). Reducing the racial achievement gap: A social-psychological intervention. *Science*, *313*(5791), 1307–1310.

Dweck, C. S. (2006). *Mindset: The new psychology of success.* Random House.

Ramirez, G., & Beilock, S. L. (2011). Writing about testing worries boosts exam performance in the classroom. *Science*, *331*(6014), 211–213.

Steele, C. M., & Aronson, J. (1995). Stereotype threat and the intellectual test performance of African Americans. *Journal of Personality and Social Psychology*, *69*(5), 797–811.

Supplemental Features for This Chapter

Math Mindset Moment: a short, boxed feature asks students to write their own three-sentence values affirmation, ready to reread the night before their next math test. The feature closes with the line: "Fear may knock. Confidence answers the door."

Student Success Story: a composite profile follows a student who had frozen on every timed math test since middle school, until a ten-minute expressive-writing habit before each test, paired with a rehearsed fearless-response routine, replaced panic with a plan. The story traces the specific exam when that student, for the first time, finished a timed test with minutes to spare instead of racing an unbeaten clock.

Reflection Questions:

(a) write your own three-sentence values-affirmation to reread before your next math test;

(b) describe one moment when test anxiety got in the way of clear thinking, and how expressive writing might have helped;

(c) explain, in your own words, why stereotype threat is situational and temporary rather than a fixed limitation;

(d) write your own four-step fearless-response routine for the moment a hard problem first appears;

(e) describe what walking into your next test with a full plan, instead of hope alone, would look like.

Chapter Takeaway

Confidence traces back to evidence and practice, not to a trait a student either has or lacks from birth (Bandura, 1977). Brief, rehearsed exercises, a values-affirmation, expressive writing before a test, and a fearless-response routine for the first hard problem, each carry a measurable, documented effect on performance (Cohen et al., 2006; Ramirez & Beilock, 2011). Stereotype threat operates as a situational pressure a student can name and set aside, not a verdict on that student's actual ability (Steele & Aronson, 1995). Every mindset practiced since Chapter 2 and every system built since Chapter 4 now has a final layer added on top: a specific plan for walking into any room, any test, any hard problem with belief rather than dread (Dweck, 2006). Chapter 15 closes the book with a complete blueprint, drawing every strategy from these fourteen chapters into one lasting plan for a student's future in math and beyond.

Chapter 15: Conquer Math. Conquer Fear.

The Blueprint for Lifelong Success

Getting Started

The journey ahead has never been solely about learning mathematics; it has always been about discovering the confidence, perseverance, and resilience that already exist within you. Mathematics is simply the vehicle that teaches these lifelong qualities. Every challenge you have overcome, every mistake you have corrected, and every concept you have mastered has prepared you for much more than a math test, ready to help you face life's challenges with courage and determination.

As you move forward, remember that mathematics is present in every stage of life. Whether you are buying your first home, investing for retirement, starting a business, managing a family budget, planning a vacation, or pursuing a college degree, mathematical thinking will continue to guide your decisions. Imagine purchasing your first home. You'll compare mortgage rates, calculate monthly payments, estimate property taxes, and develop a long-term financial plan. The confidence you developed while solving math problems in school becomes the confidence you use to build a successful future.

Never allow fear to convince you that you are incapable of greatness. You are stronger than your doubts, smarter than your mistakes, and more resilient than any obstacle you may face. Conquering mathematics comes down to conquering the fear that limits your potential. As you continue your educational journey, remember this simple truth: every problem has a solution, every challenge is an opportunity to grow, and every student has the ability to succeed. Believe in yourself, embrace the journey, and never stop learning. When you conquer your fear, you don't just conquer math: you unlock a future filled with limitless possibilities.

Real-Life Math Connection

Buying Your First Home

Solution

Home Price

$320,000

20% Down Payment

320,000(.20)

=

$64,000

Mortgage

320,000 − 64,000

=

$256,000

Monthly mortgage payments are calculated using interest rates, loan terms, taxes, and insurance. Understanding mathematics empowers future homeowners to make informed financial decisions and avoid costly mistakes.

Chapter Purpose and Guiding Question

What happens the day a student closes this book and carries its strategies into an ordinary math class?

Fourteen chapters have built a full toolkit, mindset, vocabulary, organization, reasoning, focus, error analysis, identity, teacher and family support, discipline, purpose, and confidence, and a toolkit only matters once a student carries it somewhere the book cannot follow. Perseverance and passion for a long-term goal predicted achievement across settings that looked nothing alike, from a spelling bee to a military academy to an Ivy League classroom, and that trait mattered more than talent alone in every one of those settings (Duckworth et al., 2007). A brief values-affirmation exercise produced gains in grades that lasted for years after the exercise ended, proof that a short, well-placed practice can carry weight far beyond the day a student first tries it (Cohen, Garcia, Apfel, & Master, 2006). A belief that ability grows with effort, once it takes root, continues to shape a student's choices long after any single class has ended (Dweck, 2006). Chapter 15 draws these findings into one closing argument: the strategies in this book work because they build habits, not because they solve any single test, and habits outlast the classroom that first taught them. The remaining sections give students a blueprint for carrying forward every earlier chapter, one deliberate choice at a time, into a future this book will never see.

Learning Objectives

Chapter 15 sets five learning outcomes for students to reach before finishing this book:

(a) explain why grit, perseverance, and passion sustained toward a long-term goal predict achievement beyond raw talent (Duckworth et al., 2007);

(b) identify three strategies from earlier chapters the student has already practiced with success;

(c) write a personal math mission statement connecting daily effort to a long-term goal;

(d) describe how a growth mindset extends beyond math class into other areas of life (Dweck, 2006);

(e) create a one-page personal blueprint listing specific habits the student commits to carrying forward.

Section Outline

I. The Toolkit Recap: Everything This Book Has Built

A student who has read all fourteen prior chapters now holds tools most students never receive in one place: a mindset for treating failure as data, a vocabulary for naming confusion out loud, a system for organizing notes and time, a habit of writing out reasoning, a routine for regaining focus, a method for learning from errors, a sense of identity strong enough to hold up under pressure, a circle of support at school and at home, a disciplined practice schedule, a set of real-world reasons math matters, and a rehearsed plan for walking into any test with belief instead of dread. No single tool on this list carries the full weight of a student's math future; the tools work as a set, each covering a gap the others leave open. Chapter 15 asks students to name, in their own words, three tools from this list that they have already used successfully, turning an abstract recap into a personal inventory.

(a) A full toolkit spanning mindset, systems, reasoning, support, discipline, purpose, and confidence;

(b) no single strategy carrying the whole weight of a student's math future on its own;

(c) a personal inventory listing three tools the student has already used successfully.

II. Grit: The Trait That Outlasts Any Single Class

Talent explains why a problem feels easy on a given day; grit explains why a student keeps working on a problem that does not feel easy at all. Perseverance and passion sustained toward a long-term goal predicted success across settings as different as a national spelling bee, a demanding military academy, and an Ivy League classroom, and this trait added predictive power beyond what talent measures alone could explain (Duckworth et al., 2007). Grit does not mean never struggling; it means returning to the same long-term goal the next day, the next week, and the next semester, even after a setback. A student who has practiced every strategy in this book has also been practicing grit itself, one hard problem at a time, without needing a name for it until now.

(a) Grit is defined as perseverance and passion sustained toward a long-term goal;

(b) grit's predictive power across varied high-stakes settings, beyond talent alone (Duckworth et al., 2007);

(c) grit built without announcement through every earlier chapter's practice, long before this chapter names it.

III. Habits That Outlast the Classroom

A strategy practiced once helps for a day; a strategy practiced until it becomes a habit helps for a lifetime, since a habit runs without requiring a student to remember why it matters. A brief values-affirmation exercise, practiced once, produced grade gains that lasted for years afterward, evidence that a short, well-placed habit can carry weight far past the moment a student first tries it (Cohen et al., 2006). Chapter 15 asks students to choose the three strategies from this book most worth turning into permanent habits, whether that means a weekly study schedule, a self-affirmation before a stressful task, or a fearless-response routine for a hard problem. A habit chosen with intention now saves a student from needing to relearn this entire book in a future moment of difficulty.

(a) The difference between a strategy tried once and a strategy turned into a lasting habit;

(b) a documented example of a brief practice producing effects that lasted for years (Cohen et al., 2006);

(c) three strategies chosen with intention to become permanent, ongoing habits.

IV. Beyond Math: A Growth Mindset for Life

A growth mindset, once built in a math classroom, does not stay contained inside that classroom; it reshapes how a person approaches any hard skill, from a musical instrument to a new language to a first job. A belief that ability grows through effort and strategy, rather than arriving fixed at birth, shaped how people responded to setbacks across school, work, and relationships, not only inside a single subject (Dweck, 2006). A student who has learned to treat a wrong answer as useful information in math class carries that same skill into a difficult conversation, a missed opportunity, or a new challenge years from now. Chapter 15 asks students to name one area of life outside math in which they plan to apply the same growth-oriented response to difficulty.

(a) A growth mindset extending past math class into any hard skill a person pursues (Dweck, 2006);

(b) treating a wrong answer, a setback, or a missed opportunity as useful information rather than a verdict;

(c) one specific area outside math where a student commits to applying this same mindset.

V. Writing a Personal Math Mission Statement

A mission statement turns a vague hope into a specific commitment a student can return to on a hard day, the same way a written goal outperformed a vague intention in Chapter 12. A strong mission statement names a long-term goal, connects that goal to a daily habit, and states, in the student's own words, why the effort is worth sustaining even when it feels hard. Chapter 13 gave students the language to connect math to a career or a business idea; Chapter 15 asks students to fold that language into one short paragraph they can read again next month, next year, and beyond. A mission statement does not need to predict a student's exact future career; it only needs to give that student a reason to keep showing up for math practice on an ordinary Tuesday.

(a) A mission statement naming a long-term goal, a daily habit, and a personal reason for the effort;

(b) written goals outperforming vague intentions, echoing the goal-setting research from Chapter 12;

(c) a mission statement built to be reread next month, next year, and beyond, not filed away and forgotten.

VI. The Blueprint: A One-Page Plan to Carry Forward

Chapter 15 closes with a single deliverable: a one-page personal blueprint that brings together every earlier chapter into a document a student can keep, post, or return to whenever math starts to feel like a threat instead of a subject. The blueprint names the three habits a student commits to carrying forward, the personal mission statement written in Section V, the fearless-response routine built in Chapter 14, and one name, a teacher, family member, or mentor, the student can turn to when a setback arrives. A student who completes this blueprint has not finished learning math; that student has finished building the system that will carry them through everything still ahead. The book ends here, yet the student's math story, built across fifteen chapters one page at a time, is only getting started.

(a) A one-page blueprint combining committed habits, a mission statement, and a fearless-response routine;

(b) one named person the student can turn to for support when a setback arrives;

(c) the blueprint framed as a beginning, not a conclusion, to the student's math story.

Research Foundations Cited in This Chapter

Cohen, G. L., Garcia, J., Apfel, N., & Master, A. (2006). Reducing the racial achievement gap: A social-psychological intervention. *Science*, *313*(5791), 1307–1310.

Duckworth, A. L., Peterson, C., Matthews, M. D., & Kelly, D. R. (2007). Grit: Perseverance and passion for long-term goals. *Journal of Personality and Social Psychology*, *92*(6), 1087–1101.

Dweck, C. S. (2006). *Mindset: The new psychology of success.* Random House.

Supplemental Features for This Chapter

Math Mindset Moment: a closing boxed feature asks students to write their finished blueprint on a single index card and keep that card somewhere visible, a binder, a mirror, a phone case, for the rest of the school year. The feature closes with the line: "The book ends. My story does not."

Student Success Story: a composite profile follows a student across all fifteen chapters, from a first day spent in silence and avoidance to a final semester where that student led a study group, mentored a younger sibling through fractions, and named engineering as a real, chosen goal. The story closes with that student's own one-page blueprint, written in their own words, offered as a model for the blueprint every reader is about to write.

Reflection Questions:

(a) name three strategies from this book you have already used with success, and describe one moment each one helped;

(b) write your own personal math mission statement, naming a long-term goal and a daily habit connected to it;

(c) describe one area of life outside math where you plan to apply a growth mindset going forward;

(d) name one teacher, family member, or mentor you can turn to when a future setback arrives;

(e) complete your one-page personal blueprint and choose one visible place to keep it for the rest of the year.

Chapter Takeaway: The Book's Closing Argument

Grit, perseverance, and passion sustained toward a long-term goal predicted achievement across some of the most demanding settings researchers have studied, and it predicted that achievement beyond what raw talent alone could explain (Duckworth et al., 2007). A brief, well-placed practice produced effects that lasted for years, proof that a habit built with intention outlives the moment a student first builds it (Cohen et al., 2006). A growth mindset, once cultivated in a math classroom, accompanies a student into every challenging skill still ahead of them, both in school and far beyond it (Dweck, 2006). Fear does not disappear the day a student closes this book, and it was never the goal of this book to make fear disappear; the goal was to build a student capable of moving forward with fear present instead of in charge. Conquer math. Conquer fear. The blueprint is written. The rest of the story belongs to the student holding it.

Teacher's Guide

Classroom Implementation Notes for All Fifteen Chapters

Purpose of This Guide

Each chapter of the book becomes, in this guide, a specific set of classroom moves a teacher can put into practice starting the same week that chapter gets assigned. Chapter 10 named five research-grounded teacher practices: high expectations paired with support, culturally responsive teaching, trust-building, instructional clarity, and differentiated instruction, and each entry in this guide applies at least one of those five practices to the content of its own chapter (Gay, 2000; Hamre & Pianta, 2001; Ladson-Billings, 1995; Tomlinson, 2001). Teacher expectations shape student outcomes through a documented, measurable mechanism, meaning the language and structure a teacher brings to each chapter carry real weight beyond the words on the page (Rosenthal & Jacobson, 1968). A teacher does not need to complete every suggestion in this guide to see a benefit; even one consistent practice, applied across a full term, builds toward the kind of classroom culture Chapter 10 describes. Each entry below names the chapter, restates its core classroom purpose in a single sentence, and offers three concrete implementation moves a teacher can adapt to a specific classroom and student population.

How to Use This Guide

Assign chapters in order, since each chapter builds on strategies introduced earlier, and the book's own transition paragraphs are written to carry that continuity forward. A pace of about one chapter every one to two weeks gives students enough time to practice a chapter's core strategy before the next chapter introduces something new. Reflection Questions at the end of each chapter work well as an exit ticket, a warm-up the following day, or a small-group discussion starter, depending on the time available in a given class period.

(a) Assign chapters in sequence rather than skipping ahead, since later chapters reference strategies built in earlier ones;

(b) pair each chapter's Reflection Questions with a short class discussion, giving students a chance to hear how classmates answered the same prompt;

(c) revisit the Math Mindset Moment quote from a completed chapter at the start of a later class period as a quick recall and motivation check.

Classroom Notes for Every Chapter

Chapter 1: Math Is Not the Enemy

Chapter 1 works best as a shared classroom experience early in the term, since it gives students shared language for a fear cycle many of them have never heard named out loud. A teacher who reads the chapter's description of the mask of indifference aloud, then asks whether anyone recognizes that pattern in themselves, often finds more honesty in the room than a written prompt alone would produce. Naming stereotype threat in plain, age-appropriate terms follows the same logic that made naming it effective in Chapter 14 (Steele & Aronson, 1995).

(a) Open the term with a class discussion built around the mask-of-indifference concept, inviting students to share without penalty;

(b) post the chapter's growth-mindset language next to the classroom door as a daily reminder on the way in;

(c) use the Student Success Story as a model, then invite students to draft one honest paragraph about their own math history.

Chapter 2: Rewriting the Narrative

A teacher's own language shapes whether a classroom sounds like a fixed-mindset room or a growth-mindset room, which makes this chapter as much a teacher exercise as a student one. Modeling the specific phrase swaps this chapter teaches, replacing "I'm not a math person" with "I haven't learned this yet," gives students a live example every time a teacher narrates their own thinking at the board. The brain-based evidence behind this chapter, showing that structured practice reshapes the brain at any age, gives skeptical students a concrete, physical reason to trust the process (Maguire et al., 2000).

(a) Post a visible list of fixed-mindset phrases next to their growth-mindset replacements, and refer to it out loud during instruction;

(b) invite students to write and revise their own personal math narrative at the start and again near the end of the term;

(c) narrate a teacher's own productive struggle with an unfamiliar problem in front of the class at least once during this unit.

Chapter 3: Math Has a Language

Vocabulary instruction pays off across every later chapter, since a student who cannot name a concept struggles to reason about it out loud or in writing. A weekly Frayer model routine, built into a consistent day and time, turns vocabulary building into a habit rather than a one-time lesson (Frayer et al., 1969). A classroom word wall, updated as new terms appear, provides a student with a reference tool during independent work without requiring the student to ask for help in front of peers.

(a) Dedicate one class period per week to a Frayer model routine for a new or difficult term;

(b) build and maintain a classroom word wall, adding a term the same day it first appears in a lesson;

(c) flag words with double meanings in everyday speech, such as "table" or "root," the first time each one appears in a math context.

Chapter 4: Organization Is a Math Superpower

Organization functions as an executive function skill rather than a fixed trait, which means a teacher can build it through explicit instruction rather than assuming a student either has it or lacks it (Dawson & Guare, 2010). Modeling a binder or notebook setup on the first day of this unit, rather than describing it in words alone, gives students a physical template to copy right away. A brief weekly binder or notebook check, kept low-stakes and focused on completion rather than perfection, reinforces the habit without turning organization into another source of anxiety.

(a) Model a complete notebook or binder setup on the board during the first class period of this unit;

(b) provide a printed graphic organizer template for the current unit rather than asking students to build one from scratch;

(c) run a brief, low-stakes organization check once a week, focused on completion rather than neatness.

Chapter 5: The Power of Note-Taking

A math journal works only if a teacher sets aside time for it, since a strategy competing with a packed lesson plan tends to fall away within the first two weeks. Five minutes at the start or end of class, used on a fixed schedule for a brief written reflection on a recent mistake, builds the metacognitive habit described in this chapter without requiring a separate lesson (Flavell, 1979). Annotating one worked example together as a whole class, using the same annotation marks the book introduces, gives students a shared visual language to carry into independent work.

(a) Protect five minutes at the start or end of class for a written math-journal reflection, on a fixed and predictable schedule;

(b) model annotation on one worked example per unit using a document camera or projected screen;

(c) collect journals at set points in the term for a completion check rather than a graded review, preserving the honesty of student reflections.

Chapter 6: Learning the Process, Not Just the Answer

A classroom built around answer-getting trains a different habit than a classroom built around sense-making, and the shift between the two often comes down to the specific questions a teacher asks after a student answers. Asking "how do you know" instead of moving straight to the next problem gives every student practice with the self-explanation effect this chapter describes (Chi et al., 1989). Structured math-talk moves, such as asking one student to restate a classmate's reasoning, build the kind of mathematical discourse this chapter treats as a core classroom practice (Chapin et al., 2003).

(a) Ask "how do you know" after a correct answer at least as often as after an incorrect one;

(b) introduce two or three specific math-talk moves and use consistent, named language for each one;

(c) structure at least one peer-collaboration task per unit around a written or verbal explanation, not only a shared answer.

Chapter 7: Focus Over Frustration

A student whose attention drifts under the pull of anxiety needs a reset routine, not a reminder to concentrate, since anxious attention competes for the exact mental resources concentration requires (Ashcraft, 2002). A short, consistent focus-building routine at the start of a lesson, even a single minute of quiet breathing, trains attention the same way a warm-up trains a muscle before harder work begins (Mrazek et al., 2013). A brief mid-lesson pause, offered before frustration peaks rather than after, keeps a difficult problem set from becoming a lost class period.

(a) Open each lesson in this unit with the same one-minute focus routine, repeated until it becomes automatic;

(b) build a scheduled mid-lesson pause into any problem set expected to run longer than fifteen minutes;

(c) teach the chapter's mindfulness strategy in clear, guided steps, then practice it together before asking students to try it on their own.

Chapter 8: Mistakes Are Part of Mastery

A classroom that punishes visible mistakes teaches students to hide their thinking, while a classroom that treats a mistake as data teaches students to keep reasoning out loud even when they are unsure. Public error analysis, performed on an anonymous or volunteered mistake rather than a single student's paper, builds the classroom norm this chapter asks teachers to create. Feedback that names a specific next step, rather than a general judgment, produced stronger learning gains than feedback focused only on whether an answer was right or wrong (Hattie & Timperley, 2007).

(a) Select one anonymous or volunteered mistake per week for a whole-class error-analysis discussion;

(b) replace general feedback such as "not quite" with feedback naming one specific next step;

(c) state a classroom norm out loud, such as "a wrong answer here is useful information," and repeat it as a standing rule.

Chapter 9: Peer Pressure and Performance

A classroom culture where visible intelligence carries a social cost will not shift through a single lesson, but a teacher who names that cost in plain terms, the way this chapter does, gives students language for a pressure many have felt without a name for it. Featuring mathematicians, scientists, and engineers from a range of backgrounds as a routine, year-round part of instruction, not a single unit, builds the kind of representation this chapter treats as a resource rather than a bonus. A brief belonging exercise, reminding students that early struggle is common and temporary, produced measurable academic gains that lasted for years (Walton & Cohen, 2011).

(a) Feature a mathematician, scientist, or engineer from a range of backgrounds in a warm-up or word problem at least once every unit, not once a year;

(b) run a short belonging exercise near the start of a new or difficult unit, normalizing early struggle as common and temporary;

(c) identify and support one student ready to model academic leadership, such as answering first or defending a correct but unpopular answer.

Chapter 10: Teachers Who Transform Math Culture

Chapter 10 speaks to the teacher, not the student, which makes it the ideal chapter to reread on its own partway through a term as a self-check rather than only a one-time read. The five practices this chapter names, expectations paired with support, cultural responsiveness, trust, clarity, and differentiation, reinforce one another, so strengthening any single practice tends to strengthen the rest (Gay, 2000; Hamre & Pianta, 2001; Ladson-Billings, 1995; Rosenthal & Jacobson, 1968; Tomlinson, 2001). A teacher returning to this chapter mid-term might choose the one practice feeling weakest that month and focus improvement there rather than attempting all five at once.

(a) Reread this chapter at the midpoint of a term and choose one practice to strengthen on purpose for the remaining weeks;

(b) pair a high-expectation statement with a specific, named form of support every time one is given to a struggling student;

(c) design one tiered task this month offering at least two entry points into the same core concept.

Chapter 11: Families, Mentors, and Math Success

A teacher cannot control what happens at home, but a teacher can make it easier for families to support a student's education, since families involved in a student's education produced stronger achievement when that involvement connected to specific classroom learning goals (Henderson & Mapp, 2002). A short, regular note home naming one specific, non-technical question a parent could ask, such as "what is one thing you learned in math today," removes the barrier many families feel around helping with math they do not remember well themselves. Coordinating with the strategies in this chapter, rather than assuming families already know them, extends a teacher's classroom work into hours the classroom cannot reach.

(a) Send home one specific, non-technical question each week that a family member can ask without needing math expertise;

(b) invite families to a brief math night built around one strategy from this book, rather than a general open house;

(c) send this chapter's guidance on process-focused praise home to families through a newsletter or class communication.

Chapter 12: The Mathematics of Discipline

Discipline systems work better when a teacher builds them into classroom structure rather than leaving them to each student's individual habits. Spaced, short practice sessions produced stronger retention than a single long session, which supports building brief, frequent practice into daily routine rather than saving all practice for the night before a test (Cepeda et al., 2006). Specific, written goals, paired with a plan and regular feedback, outperformed vague goals such as "do better," giving a teacher a concrete structure to require rather than only suggest (Locke & Latham, 2002).

(a) Build ten to fifteen minutes of spaced practice into most class periods rather than relying on occasional long review sessions;

(b) require a written, specific goal from each student before a unit test, including a number and a plan;

(c) pair students as accountability partners for a defined period and check in on that partnership for a few minutes each week.

Chapter 13: Real-Life Math and Future Opportunities

Relevance functions as its own motivational lever, separate from confidence, since a brief exercise connecting course material to a student's own life raised both interest and grades over a full semester (Hulleman & Harackiewicz, 2009). Inviting a guest speaker from a STEM career, a trade, or a local business gives students a firsthand look at the connections this chapter describes in writing. Framing a unit's word problems around real financial, entrepreneurial, or career scenarios turns an abstract skill into a tool students can see themselves using outside the classroom.

(a) Invite one guest speaker per term from a STEM career, a trade, or a small business connected to the current unit;

(b) rewrite at least one homework problem per unit around a real financial, entrepreneurial, or career scenario;

(c) assign the chapter's relevance-writing exercise as a graded but low-stakes reflection near the start of a new unit.

Chapter 14: Building the Confident Math Student

Test-day rituals matter as much as test-day content, since a brief writing exercise completed right before an exam raised scores among the most anxious test-takers in a room (Ramirez & Beilock, 2011). Building two or three minutes of quiet writing time into the start of a major test, framed as the chapter's expressive-writing exercise, costs little instructional time while targeting the anxiety this chapter addresses. Naming stereotype threat out loud before a high-stakes test, in age-appropriate terms, has itself been shown to reduce its effect on performance (Steele & Aronson, 1995).

(a) Build two to three minutes of expressive-writing time into the start of every major test this term;

(b) teach the chapter's fearless-response routine in clear, practiced steps, then practice it together on a problem chosen for its difficulty before test day;

(c) offer students a brief reminder, before a high-stakes test, that any extra pressure they may feel is temporary and situational.

Chapter 15: Conquer Math. Conquer Fear.

Chapter 15 works well as a capstone project rather than a standard reading assignment, giving students a reason to look back over the full term rather than only forward to the next unit. Asking students to present their one-page personal blueprint to a small group, or to a family member at home, turns a private planning document into a spoken commitment, which tends to strengthen follow-through. Perseverance and passion sustained toward a long-term goal predicted achievement beyond talent alone, giving a teacher a research-grounded reason to treat this closing project as more than a formality (Duckworth et al., 2007).

(a) Assign the personal blueprint as a small capstone project, with class time set aside to complete and share it;

(b) invite students to present one committed habit from their blueprint to a small group or a family member;

(c) keep a copy of each student's blueprint on file, with permission, to return to that student at the start of the following year.

Research Foundations Cited in the Teacher's Guide

Ashcraft, M. H. (2002). Math anxiety: Personal, educational, and cognitive consequences. *Current Directions in Psychological Science*, *11*(5), 181–185.

Cepeda, N. J., Pashler, H., Vul, E., Wixted, J. T., & Rohrer, D. (2006). Distributed practice in verbal recall tasks: A review and quantitative synthesis. *Psychological Bulletin*, *132*(3), 354–380.

Chapin, S. H., O'Connor, C., & Anderson, N. C. (2003). *Classroom discussions: Using math talk to help students learn*. Math Solutions Publications.

Chi, M. T. H., Bassok, M., Lewis, M. W., Reimann, P., & Glaser, R. (1989). Self-explanations: How students study and use examples in learning to solve problems. *Cognitive Science*, *13*(2), 145–182.

Dawson, P., & Guare, R. (2010). *Executive skills in children and adolescents: A practical guide to assessment and intervention* (2nd ed.). The Guilford Press.

Duckworth, A. L., Peterson, C., Matthews, M. D., & Kelly, D. R. (2007). Grit: Perseverance and passion for long-term goals. *Journal of Personality and Social Psychology*, *92*(6), 1087–1101.

Flavell, J. H. (1979). Metacognition and cognitive monitoring: A new area of cognitive-developmental inquiry. *American Psychologist*, *34*(10), 906–911.

Frayer, D. A., Frederick, W. C., & Klausmeier, H. J. (1969). *A schema for testing the level of concept mastery* (Technical Report No. 16). University of Wisconsin, Research and Development Center for Cognitive Learning.

Gay, G. (2000). *Culturally responsive teaching: Theory, research, and practice.* Teachers College Press.

Hamre, B. K., & Pianta, R. C. (2001). Early teacher-child relationships and the trajectory of children's school outcomes through eighth grade. *Child Development*, *72*(2), 625–638.

Hattie, J., & Timperley, H. (2007). The power of feedback. *Review of Educational Research*, *77*(1), 81–112.

Henderson, A. T., & Mapp, K. L. (2002). *A new wave of evidence: The impact of school, family, and community connections on student achievement*. Southwest Educational Development Laboratory.

Hulleman, C. S., & Harackiewicz, J. M. (2009). Promoting interest and performance in high school science classes. *Science*, *326*(5958), 1410–1412.

Ladson-Billings, G. (1995). Toward a theory of culturally relevant pedagogy. *American Educational Research Journal*, *32*(3), 465–491.

Locke, E. A., & Latham, G. P. (2002). Building a practically useful theory of goal setting and task motivation: A 35-year odyssey. *American Psychologist*, *57*(9), 705–717.

Maguire, E. A., Gadian, D. G., Johnsrude, I. S., Good, C. D., Ashburner, J., Frackowiak, R. S., & Frith, C. D. (2000). Navigation-related structural change in the hippocampi of taxi drivers. *Proceedings of the National Academy of Sciences*, *97*(8), 4398–4403.

Mrazek, M. D., Franklin, M. S., Phillips, D. T., Baird, B., & Schooler, J. W. (2013). Mindfulness training improves working memory capacity and GRE performance while reducing mind wandering. *Psychological Science*, *24*(5), 776–781.

Ramirez, G., & Beilock, S. L. (2011). Writing about testing worries boosts exam performance in the classroom. *Science*, *331*(6014), 211–213.

Rosenthal, R., & Jacobson, L. (1968). *Pygmalion in the classroom: Teacher expectation and pupils' intellectual development*. Holt, Rinehart & Winston.

Steele, C. M., & Aronson, J. (1995). Stereotype threat and the intellectual test performance of African Americans. *Journal of Personality and Social Psychology*, *69*(5), 797–811.

Tomlinson, C. A. (2001). *How to differentiate instruction in mixed-ability classrooms* (2nd ed.). Association for Supervision and Curriculum Development.

Walton, G. M., & Cohen, G. L. (2011). A brief social-belonging intervention improves academic and health outcomes of minority students. *Science*, *331*(6023), 1447–1451.

Closing Note for Teachers

None of the fifteen chapters in this book expects a teacher to work alone; each one names a specific, learnable practice that a teacher can build one week at a time, just as the book asks students to build their own strategies one chapter at a time. A classroom culture shifts through hundreds of small, consistent choices rather than a single dramatic lesson, and this guide aims to make those choices concrete rather than abstract (Ladson-Billings, 1995). A teacher who picks even three or four suggestions from this guide and applies them consistently across a full term gives students a classroom where the book's message and the room's daily practice say the same thing.

Family and Mentor Guide

Supporting a Student Through All Fifteen Chapters

Purpose of This Guide

Parents, guardians, and mentors gain a specific, low-barrier way to support a student working through this book, without needing to remember a single math procedure. Family involvement supported achievement across a wide range of studies, and the strongest results appeared when that involvement connected to specific learning goals rather than general encouragement alone (Henderson & Mapp, 2002). A parent's own math anxiety can be passed to a child through anxious help, an effect documented among parents who reported frequently providing math-anxious help with homework; thus, the goal of this guide is never to turn a family member into a math tutor (Maloney et al., 2015). Praise aimed at a student's effort and strategy built stronger persistence after a setback than praise aimed at fixed intelligence, a small language shift any family member can practice starting today (Mueller & Dweck, 1998). Community mentoring extended a family's support beyond what any single household could offer alone, adding consistent, caring adults to a student's circle of support (DuBois et al., 2011). Each entry below names a chapter and offers three specific things a family member or mentor can do at home, none of which require solving a single math problem.

How to Use This Guide

Read the chapter a student is working through at the moment, or read the short summary below if time is limited, then choose one suggestion rather than attempting all three at once. A consistent small habit, repeated across a full term, builds more trust and confidence than an occasional large gesture. Chapter 11 offers the deepest treatment of family and mentor support in the book and rewards a full read on its own, alongside this guide.

(a) Pick one suggestion per chapter rather than all three, and repeat it until it becomes a habit;

(b) ask a specific question about the chapter's content, such as "what is one thing you practiced this week," rather than a general question like "how was math";

(c) revisit Chapter 11 whenever a home strategy stops working, since it offers the most comprehensive treatment of family support in the book.

Home Notes for Every Chapter

Chapter 1: Math Is Not the Enemy

Chapter 1 names a fear cycle many students carry in silence, and a family member's own honest story about math, shared without apology, often opens the door to a student's own story in return. A family member who once struggled with math can say so out loud without undermining a student's confidence, provided that admission comes paired with belief in the student's own future (Steele & Aronson, 1995).

(a) Share one honest sentence about a family member's own math history, paired with one sentence of belief in the student's future;

(b) ask whether the student has ever hidden a correct answer or slowed down on purpose to avoid standing out, and listen without judgment;

(c) avoid saying "I was never a math person" in front of the student, even as a joke.

Chapter 2: Rewriting the Narrative

The language a family uses about math shapes a student's own internal language, which makes this chapter as much a household practice as a classroom one. Swapping "you're so smart" for "you worked hard on that" at home builds the same growth-oriented habit this chapter teaches in class (Dweck, 2006).

(a) Replace "you're so smart" with specific praise for effort, strategy, or persistence;

(b) ask the student to explain one thing they used to believe about their own math ability that no longer feels true;

(c) share one thing a family member is learning right now as an adult, modeling growth at any age.

Chapter 3: Math Has a Language

A family member does not need to know a single math term to help build vocabulary; they only need to ask a student to explain a new word in their own words. Asking "What does that word mean in math class?" gives a student low-stakes practice with the exact skill this chapter builds.

(a) Ask the student to teach one new vocabulary word from class each week, in their own words;

(b) post the student's own vocabulary list somewhere visible at home, such as the refrigerator;

(c) notice out loud when a math word sounds like an everyday word with a different meaning, such as "table" or "volume."

Chapter 4: Organization Is a Math Superpower

A quiet, consistent homework space, kept stocked with basic supplies, supports the organizational systems this chapter teaches without requiring a family member to understand the content inside a student's binder. A brief, friendly check of whether a binder or notebook is up to date, kept low-stakes, reinforces the habit without adding pressure.

(a) Set up one consistent, quiet space at home for math homework and studying;

(b) ask to see the student's math binder or notebook once a week, without checking for correctness;

(c) keep basic supplies, folders, and paper stocked and easy to find.

Chapter 5: The Power of Note-Taking

A family member can ask to hear about the student's math journal without needing to read a single equation inside it, since the point of the question is the reflection, not the content. Asking "what mistake did you learn from this week" gives a student practice putting a setback into words out loud, a skill this chapter builds through writing alone.

(a) Ask the student to share one entry from their math journal, focused on the reflection rather than the math itself;

(b) ask "what mistake did you learn from this week" as a regular dinner-table or car-ride question;

(c) celebrate a shared mistake and its lesson the same way a family might celebrate a correct answer.

Chapter 6: Learning the Process, Not Just the Answer

Asking a student to explain their reasoning out loud gives a family member real insight into a student's thinking, even when that family member could not solve the problem themselves. "Walk me through how you got that" works as well at a kitchen table as it does in a math classroom (Chi et al., 1989).

(a) Ask "walk me through how you got that" instead of only checking whether an answer is correct;

(b) let the student teach a family member a concept from class, reversing the usual roles;

(c) resist the urge to provide an answer, even when a family member knows it.

Chapter 7: Focus Over Frustration

A calm homework environment, free of a loud television or constant interruption, supports the focus-building routines this chapter teaches without requiring a family member to teach a single strategy. A brief, shared breathing routine practiced together before homework begins gives a student one more place to rehearse a skill from class.

(a) Reduce noise and interruption during a set homework window each day;

(b) practice the chapter's brief breathing routine together before homework begins, at least once;

(c) notice and name frustration in a calm voice, such as "this looks like a hard one," instead of rushing a student past it.

Chapter 8: Mistakes Are Part of Mastery

A family member's reaction to a low grade or a wrong answer teaches a student more than any lecture about trying harder ever could. Responding to a mistake with curiosity, such as "what do you think happened here," instead of frustration or punishment, extends the classroom error culture this chapter teaches into the home (Moser et al., 2011).

(a) Respond to a low grade or a mistake with curiosity first, before any conversation about consequences;

(b) ask "what do you think happened here" rather than "why did you get this wrong";

(c) share a mistake a family member made not long ago, along with what that mistake taught them.

Chapter 9: Peer Pressure and Performance

A family can extend the representation this chapter treats as a resource, since a student who sees mathematicians, scientists, and engineers from a range of backgrounds named at home, not only in a classroom, builds a fuller picture of who belongs in those fields. Celebrating a moment of visible academic effort at home, rather than treating it as unremarkable, pushes back against the social cost this chapter describes.

(a) Name one mathematician, scientist, or engineer at home and talk about their path into the field;

(b) celebrate a moment when the student answered with confidence or defended a correct answer;

(c) ask what it would feel like if visible effort in math carried no social cost at all, and listen to the answer.

Chapter 10: Teachers Who Transform Math Culture

A family that partners with a student's teacher, rather than working around that teacher, extends the classroom practices this chapter describes into a consistent message across school and home. A brief, friendly note or message to a teacher, asking how a family can reinforce classroom expectations at home, opens a partnership this chapter's research supports (Hamre & Pianta, 2001).

(a) Send a brief note to the teacher near the start of a term, asking how a family can reinforce classroom expectations at home;

(b) avoid contradicting a teacher's high expectations in front of the student, even during a hard week;

(c) ask the student what their teacher does that makes math class feel safe, and what that teacher does that feels hard.

Chapter 11: Families, Mentors, and Math Success

Chapter 11 offers the fullest treatment of family and mentor support in the book, and it rewards a complete read rather than a summary alone. Its core message carries through the rest of this guide: family involvement matters, a parent's own math anxiety should not become a child's anxiety, praise should target effort rather than fixed ability, and community mentors can extend support beyond any single household (DuBois et al., 2011; Henderson & Mapp, 2002; Maloney et al., 2015; Mueller & Dweck, 1998).

(a) Read this chapter in full at least once, even if time is limited for the other chapters;

(b) identify one community mentor, tutor, or program that could extend support beyond the household;

(c) choose one strategy from this chapter and commit to practicing it for a full month.

Chapter 12: The Mathematics of Discipline

A family member makes an ideal accountability partner for the goal-setting and practice systems this chapter builds, since a plan another person is watching carries more weight than a private one (Locke & Latham, 2002). A brief weekly check-in, as simple as a single question about whether a planned practice session happened, keeps a system running through a low-motivation week.

(a) Offer to serve as the student's accountability partner for one goal from this chapter;

(b) ask a brief weekly question, such as "Did you get your practice in this week?";

(c) help the student find a fixed time and place for math practice that fits the household's schedule.

Chapter 13: Real-Life Math and Future Opportunities

A family can make this chapter's message concrete through ordinary household tasks, since a budget, a grocery bill, or a recipe each turn an abstract math skill into a lived one. Involving a student in a real financial decision, even a small one, gives that student practice with the exact connection this chapter describes (Lusardi & Mitchell, 2014).

(a) Involve the student in one real household math task, such as a grocery budget or a bill;

(b) ask the student what career or business idea interests them, and connect it to a math skill from this book;

(c) share a family member's own use of math at work or in daily life, however ordinary it seems.

Chapter 14: Building the Confident Math Student

A family member can support the test-day plan this chapter outlines without needing to understand a single test question, since the plan targets confidence and calm rather than content knowledge. Reminding a student, the night before a big test, that any pressure they feel is temporary and says nothing about their actual ability echoes this chapter's core message (Steele & Aronson, 1995).

(a) Help the student write their values-affirmation sentence the night before a major test;

(b) remind the student, before a big test, that pressure is temporary and does not reflect their actual ability;

(c) avoid adding extra pressure about a test's importance in the hours leading up to it.

Chapter 15: Conquer Math. Conquer Fear.

Chapter 15 asks every student to name one adult they can turn to when a future setback arises, and a family member or mentor who says yes to that role gives a student something to hold on to long after this book closes. Asking to hear the student's finished one-page blueprint, and asking real questions about it, turns a private planning document into a shared family commitment (Duckworth et al., 2007).

(a) Ask to hear the student's finished one-page blueprint and ask at least one real question about it;

(b) agree, out loud, to be one of the people the student can turn to when a future setback arrives;

(c) keep a copy of the student's blueprint somewhere visible at home for the rest of the school year.

Research Foundations Cited in the Family and Mentor's Guide

Chi, M. T. H., Bassok, M., Lewis, M. W., Reimann, P., & Glaser, R. (1989). Self-explanations: How students study and use examples in learning to solve problems. *Cognitive Science, 13(*2), 145–182.

DuBois, D. L., Portillo, N., Rhodes, J. E., Silverthorn, N., & Valentine, J. C. (2011). How effective are mentoring programs for youth? A systematic assessment of the evidence. *Psychological Science in the Public Interest*, *12*(2), 57–91.

Duckworth, A. L., Peterson, C., Matthews, M. D., & Kelly, D. R. (2007). Grit: Perseverance and passion for long-term goals. *Journal of Personality and Social Psychology*, *92*(6), 1087–1101.

Dweck, C. S. (2006). *Mindset: The new psychology of success*. Random House.

Hamre, B. K., & Pianta, R. C. (2001). Early teacher-child relationships and the trajectory of children's school outcomes through eighth grade. *Child Development*, *72*(2), 625–638.

Henderson, A. T., & Mapp, K. L. (2002). *A new wave of evidence: The impact of school, family, and community connections on student achievement*. Southwest Educational Development Laboratory.

Locke, E. A., & Latham, G. P. (2002). Building a practically useful theory of goal setting and task motivation: A 35-year odyssey. *American Psychologist*, *57*(9), 705–717.

Lusardi, A., & Mitchell, O. S. (2014). The economic importance of financial literacy: Theory and evidence. *Journal of Economic Literature*, *52*(1), 5–44.

Maloney, E. A., Ramirez, G., Gunderson, E. A., Levine, S. C., & Beilock, S. L. (2015). Intergenerational effects of parents' math anxiety on children's math achievement and anxiety. *Psychological Science*, *26*(9), 1480–1488.

Moser, J. S., Schroder, H. S., Heeter, C., Moran, T. P., & Lee, Y. H. (2011). Mind your errors: Evidence for a neural mechanism linking growth mind-set to adaptive posterror adjustments. *Psychological Science*, *22*(11), 1484–1489.

Mueller, C. M., & Dweck, C. S. (1998). Praise for intelligence can undermine children's motivation and performance. *Journal of Personality and Social Psychology*, *75*(1), 33–52.

Steele, C. M., & Aronson, J. (1995). Stereotype threat and the intellectual test performance of African Americans. *Journal of Personality and Social Psychology*, *69*(5), 797–811.

Closing Note for Families and Mentors

None of the fifteen chapters in this book expect a family to replace a teacher, and none of the suggestions above require a family member to relearn math from the ground up. Small, repeated actions, a specific question asked at dinner, a moment of praise aimed at effort, a name shared or a story told, build the kind of trust this book depends on to work outside the classroom (Henderson & Mapp, 2002). A family or mentor who picks even one habit from this guide and keeps it going for a full term gives a student proof that the belief expressed in this book extends beyond the last page and into the home, where that student does most of their growing.

Glossary of Key Terms

A Companion Reference for All Fifteen Chapters

How to Use This Glossary

A single alphabetized reference, compiled here, collects the recurring terms introduced across the fifteen chapters of this book, so a student, teacher, or family member can consult it without searching back through earlier chapters. Each entry states a definition in plain language, then names the chapter or chapters where that term first appears in full detail. A term tied to a specific, named study lists that study in parentheses, giving a reader a direct path back to the research behind the idea. A quick, reliable check like this one serves a reader who encounters a familiar term partway through the book and wants a fast reminder of its meaning, without replacing a chapter's full explanation.

Terms A to Z

Belonging Intervention: a brief, structured exercise that helps a student understand that early struggles with belonging in a new academic setting are common and temporary, rather than a sign that the student does not belong, and has been shown to close a measurable share of an achievement gap over several years (Walton & Cohen, 2011). Discussed in Chapter 9.

Culturally Relevant Pedagogy: a teaching framework that pairs unapologetic high academic expectations with intentional respect for a student's cultural identity, treating the two goals as reinforcing rather than competing (Ladson-Billings, 1995). Discussed in Chapters 9 and 10.

Culturally Responsive Teaching: instruction filtered through a student's own cultural experiences and frames of reference, using familiar contexts, names, and communities to make new content easier to grasp and easier to trust (Gay, 2000). Discussed in Chapter 10.

Deliberate Practice: practice aimed at a specific, identified weakness, performed with full attention and paired with feedback, shown to build skill faster than unfocused repetition (Ericsson et al., 1993). Discussed in Chapter 12.

Differentiated Instruction: an instructional approach that adjusts content, process, or product in response to a student's readiness, interest, or learning profile, offering multiple entry points into the same core concept without lowering the expectation attached to it (Tomlinson, 2001). Discussed in Chapter 10.

Distributed Practice (Spaced Practice): practice spread across several separate sessions rather than crammed into one sitting, which has been shown to produce stronger long-term retention than massed practice of the same total duration (Cepeda et al., 2006). Discussed in Chapter 12.

Error Analysis: a classroom or personal practice of examining a mistake to identify its specific cause, treating the mistake as information that guides the next attempt rather than as a final verdict on ability. Discussed in Chapter 8.

Executive Function: a set of mental management skills, including planning, organizing, and holding information in mind while working, that predict math achievement and can be strengthened through structured routines (Dawson & Guare, 2010; Diamond, 2013). Discussed in Chapter 4.

Expressive Writing: a brief writing exercise, completed just before a stressful task such as an exam, in which a student writes about their worries, freeing mental resources that anxiety would otherwise consume (Ramirez & Beilock, 2011). Discussed in Chapter 14.

Fixed Mindset: a belief that intelligence and ability are stable traits a person either has or lacks, associated with avoiding challenge and giving up sooner after a setback (Dweck & Leggett, 1988). Discussed in Chapter 2.

Frayer Model: a four-part graphic organizer that asks a student to record a term's definition, its key characteristics, an example, and a non-example, used to build a deeper understanding of academic vocabulary (Frayer et al., 1969). Discussed in Chapter 3.

Grit: perseverance and passion sustained toward a long-term goal, shown to predict achievement across demanding settings beyond what talent alone could explain (Duckworth et al., 2007). Discussed in Chapter 15.

Growth Mindset: a belief that ability grows through effort, strategy, and practice rather than arriving fixed at birth, associated with greater persistence and stronger long-term achievement (Dweck, 2006). Discussed throughout the book, beginning in Chapters 1 and 2.

Math Anxiety: a feeling of tension and apprehension connected to math situations in particular, shown to interfere with working memory and measurable performance (Ashcraft, 2002). Discussed in Chapters 1 and 7.

Metacognition: the practice of thinking about one's own thinking, including monitoring understanding, evaluating a strategy, and reflecting on a mistake (Flavell, 1979). Discussed in Chapters 5 and 8.

Mindfulness (Classroom Practice): a brief attention-training exercise, such as focused breathing, used to reset a student's attention during a difficult task and shown to reduce mind wandering while supporting working memory (Mrazek et al., 2013; Zenner et al., 2014). Discussed in Chapter 7.

Process Praise: praise directed at a student's effort, strategy, or approach rather than at a fixed trait such as intelligence, shown to build stronger persistence after a later setback than praise for being smart (Mueller & Dweck, 1998). Discussed in Chapter 11.

Productive Struggle: sustained, effortful engagement with a problem that is hard but not impossible, treated as a necessary and valuable part of learning rather than a sign that something has gone wrong. Discussed in Chapters 2 and 7.

Self-Efficacy: a person's belief in their own capability to complete a specific task, built through past success, watching others succeed, encouragement, and one's own physical state, and shown to shape effort and persistence (Bandura, 1977). Discussed in Chapter 14.

Self-Explanation Effect: the finding that a student who explains their own reasoning out loud or in writing learns with greater depth than a student who only reviews a worked example (Chi et al., 1989). Discussed in Chapter 6.

Self-Regulated Learning: a cycle of planning, monitoring, and reflecting on one's own learning, repeated across a task or a term to build stronger independence and consistency (Zimmerman, 2002). Discussed in Chapter 12.

Stereotype Threat: a situational, temporary pressure that can undermine performance when a person feels at risk of confirming a negative stereotype about a group they belong to, distinct from any fixed limitation in that person's actual ability (Steele & Aronson, 1995). Discussed in Chapters 1 and 14.

Talk Moves (Mathematical Discourse): specific discussion techniques a teacher or student uses to deepen classroom conversation about math, including asking a peer to restate an idea, asking for agreement or disagreement, or asking for a next step (Chapin et al., 2003). Discussed in Chapter 6.

Think-Aloud Strategy: a technique in which a student narrates their reasoning out loud while solving a problem, making an internal thought process visible to a teacher, a peer, or the student's own later review (Davey, 1983). Discussed in Chapter 6.

Values-Affirmation (Self-Affirmation): a brief writing exercise in which a person names a personal value and explains why it matters to them, shown to reduce the psychological threat tied to a stressful academic setting and to produce achievement gains that last for years (Cohen et al., 2006). Discussed in Chapters 14 and 15.

Working Memory: the limited mental workspace a person uses to hold and manipulate information during an active task, shown to fill with anxious thoughts that compete with math reasoning for the same limited space (Ashcraft, 2002; Ramirez & Beilock, 2011). Discussed in Chapters 7 and 14.

Research Foundations Cited in This Glossary

Ashcraft, M. H. (2002). Math anxiety: Personal, educational, and cognitive consequences. *Current Directions in Psychological Science*, *11*(5), 181–185.

Bandura, A. (1977). Self-efficacy: Toward a unifying theory of behavioral change. *Psychological Review*, *84*(2), 191–215.

Cepeda, N. J., Pashler, H., Vul, E., Wixted, J. T., & Rohrer, D. (2006). Distributed practice in verbal recall tasks: A review and quantitative synthesis. *Psychological Bulletin*, *132*(3), 354–380.

Chapin, S. H., O'Connor, C., & Anderson, N. C. (2003). *Classroom discussions: Using math talk to help students learn*. Math Solutions Publications.

Chi, M. T. H., Bassok, M., Lewis, M. W., Reimann, P., & Glaser, R. (1989). Self-explanations: How students study and use examples in learning to solve problems. *Cognitive Science*, *13*(2), 145–182.

Cohen, G. L., Garcia, J., Apfel, N., & Master, A. (2006). Reducing the racial achievement gap: A social-psychological intervention. *Science*, *313*(5791), 1307–1310.

Davey, B. (1983). Think aloud: Modeling the cognitive processes of reading comprehension. *Journal of Reading*, *27*(1), 44–47.

Dawson, P., & Guare, R. (2010). *Executive skills in children and adolescents: A practical guide to assessment and intervention* (2nd ed.). The Guilford Press.

Diamond, A. (2013). Executive functions. *Annual Review of Psychology*, *64*, 135–168.

Duckworth, A. L., Peterson, C., Matthews, M. D., & Kelly, D. R. (2007). Grit: Perseverance and passion for long-term goals. *Journal of Personality and Social Psychology*, *92*(6), 1087–1101.

Dweck, C. S. (2006). *Mindset: The new psychology of success*. Random House.

Dweck, C. S., & Leggett, E. L. (1988). A social-cognitive approach to motivation and personality. *Psychological Review*, *95*(2), 256–273.

Ericsson, K. A., Krampe, R. T., & Tesch-Römer, C. (1993). The role of deliberate practice in the acquisition of expert performance. *Psychological Review*, *100*(3), 363–406.

Flavell, J. H. (1979). Metacognition and cognitive monitoring: A new area of cognitive-developmental inquiry. *American Psychologist*, *34*(10), 906–911.

Frayer, D. A., Frederick, W. C., & Klausmeier, H. J. (1969). *A schema for testing the level of concept mastery* (Technical Report No. 16). University of Wisconsin, Research and Development Center for Cognitive Learning.

Gay, G. (2000). *Culturally responsive teaching: Theory, research, and practice*. Teachers College Press.

Ladson-Billings, G. (1995). Toward a theory of culturally relevant pedagogy. *American Educational Research Journal*, *32*(3), 465–491.

Mrazek, M. D., Franklin, M. S., Phillips, D. T., Baird, B., & Schooler, J. W. (2013). Mindfulness training improves working memory capacity and GRE performance while reducing mind wandering. *Psychological Science*, *24*(5), 776–781.

Mueller, C. M., & Dweck, C. S. (1998). Praise for intelligence can undermine children's motivation and performance. *Journal of Personality and Social Psychology*, *75*(1), 33–52.

Ramirez, G., & Beilock, S. L. (2011). Writing about testing worries boosts exam performance in the classroom. *Science*, *331*(6014), 211–213.

Steele, C. M., & Aronson, J. (1995). Stereotype threat and the intellectual test performance of African Americans. *Journal of Personality and Social Psychology*, *69*(5), 797–811.

Tomlinson, C. A. (2001). *How to differentiate instruction in mixed-ability classrooms* (2nd ed.). Association for Supervision and Curriculum Development.

Walton, G. M., & Cohen, G. L. (2011). A brief social-belonging intervention improves academic and health outcomes of minority students. *Science*, *331*(6023), 1447–1451.

Zenner, C., Herrnleben-Kurz, S., & Walach, H. (2014). Mindfulness-based interventions in schools: A systematic review and meta-analysis. *Frontiers in Psychology*, *5*, Article 603.

Zimmerman, B. J. (2002). Becoming a self-regulated learner: An overview. *Theory Into Practice*, *41*(2), 64–70.

Appendix: Reproducible Templates

Five blank, photocopy-ready pages built directly from the strategies taught in this book. Print one per student, or keep a master copy for repeated use throughout the term.

Template 1: Frayer Model Vocabulary Organizer

Chapter 3, Math Has a Language. Use one page per new vocabulary term.

Term:

____

Chapter:

____

Definition (in your own words)	Characteristics or Facts
Example	**Non-Example**

Template 2: Math Journal Page

Chapter 5, The Power of Note-Taking. Adapted Cornell format for daily notes.

Name:

Date:

Topic:

Key Terms and Cues	Notes

Summary (in your own words)
One Mistake I Learned From Today

Template 3: Weekly Study Schedule

Chapter 12, The Mathematics of Discipline. Fill in a fixed time and task for each day.

Day	Time	Math Practice Task	Done?
Monday			
Tuesday			
Wednesday			
Thursday			
Friday			
Saturday			
Sunday			

Template 4: Goal Setting Card

Chapter 12, The Mathematics of Discipline. A specific goal, a deadline, and a plan; the three parts every strong goal needs.

My Goal:

____

____

Specific Number (how many, how much, what score):

____

Deadline:

____

My Plan to Get There:

____

____

Accountability Partner:

____

Check-In Day and Time:

____

Template 5: My Personal Blueprint

Chapter 15, Conquer Math. Conquer Fear. Complete this page after finishing the book and keep it somewhere visible.

Three habits I am committing to carry forward:

1.

____

2.

____

3.

____

My math mission statement:

____

____

My fearless-response routine (four steps):

1.

____

2.

____

3.

____

4.

____

One person I can turn to when a setback arrives:

Name:

____

How I will reach them:

____

About the Author

Dr. Joe Evans is an educator, instructional leader, and mathematics advocate with more than two decades of experience helping students transform math anxiety into mathematical confidence and achievement. A proud native of Baltimore, Maryland, Dr. Evans was raised by a single mother, where he learned the values of perseverance, resilience, and hard work. His passion for mathematics was ignited in high school when a teacher challenged him to solve problems at the board, teaching him that asking questions, taking academic risks, and trusting the problem-solving process are the keys to overcoming fear and building confidence. Those experiences inspired the framework behind *Conquer Math, Conquer Fear: The Focus-Driven Playbook for Turning Math Anxiety into Achievement*. He is also the author of *From the Hood to the Hood*, a memoir chronicling his journey from inner-city Baltimore to earning his doctoral hood in Educational Leadership.

A veteran of the 82nd Airborne Division, U.S. Army, Dr. Evans earned a Bachelor of Science in Mathematics from Fayetteville State University, a Master of Education in Applied Educational Technology from Wilmington University, and a Doctorate in Educational Leadership from Walden University. His educational leadership experience includes more than 18 years as a secondary mathematics teacher, over five years as a college mathematics adjunct instructor, and service as an Instructional Coach, Assistant Principal, and Principal in the state of Delaware.

As a mathematics educator, school leader, workshop presenter, entrepreneur, and co-founder of Results Tutoring, LLC, he has empowered thousands of K–12, college, and adult learners through mathematics instruction, academic interventions, leadership development, and professional learning across Delaware, neighboring states, and internationally. Dr. Evans currently serves as a Doctoral Principal Fellow at Walden University, where he continues to champion equitable learning opportunities and inspire students to replace fear with focus, confidence, and success.

Conquer Math. Conquer Fear.

by

Dr. Joe Evans

www.ingramcontent.com/pod-product-compliance
Lightning Source LLC
LaVergne TN
LVHW022215200926
848481LV00006B/1985